THE LITTLE BOOK
of TIME

THE LITTLE BOOK

of TIME

KLAUS MAINZER

Translated by Josef Eisinger

COPERNICUS BOOKS

An Imprint of Springer-Verlag

Originally published as *Zeit*,
© 1999 Verlag C. H. Beck oHG, München, Germany.

© 2002 Springer-Verlag New York, Inc.

Published in the United States by Copernicus Books,
an imprint of Springer-Verlag New York, Inc.
A member of BertelsmannSpringer Science+Business Media GmbH

Copernicus Books
37 East 7th Street
New York, NY 10003
www.copernicusbooks.com

Library of Congress Cataloging-in-Publication Data
Mainzer, Klaus.
    [Zeit. English]
    The little book of time / Klaus Mainzer.
        p.   cm.—(Little book series)
    Includes bibliographical references and index.
    ISBN 0-387-95288-8 (alk. paper)
    1. Time. I. Title. II. Little book series (New York, N.Y.)
    QB209.M3513   2002
    529—dc21   2002023357

Manufactured in the United States of America.
Printed on acid-free paper.
Translated by Josef Eisinger.

9 8 7 6 5 4 3 2 1

ISBN 0-387-95288-8   SPIN 10838269

For Johanna, Benedikt, Beatrice, Pauline, and Anna

Acknowledgments

The Little Book of Time was inspired by my research in philosophy and history of cosmology, complex systems, and nonlinear dynamics. In two books that were previously published in English, *Symmetries of Nature* (1996) and *Thinking in Complexity: The Complex Dynamics of Matter, Mind, and Mankind* (3rd edition, 1997), I explain these topics in greater detail. I want to express my gratitude to Springer-Verlag, the publisher of *Thinking in Complexity*, *The Little Book of Time*, and two of my German books, for the fruitful cooperation over the past years.

For the careful translation of *The Little Book of Time*, I'd like to thank Josef Eisinger. Thanks also to Heather Jones for largely improving the English manuscript. I am especially grateful to Mareike Paessler, Anna Painter, and Paul Farrell of Copernicus Books, as well as to designer Jordan Rosenblum, for the coordination and preparation of the book's publication.

Klaus Mainzer

Augsburg, Germany, Spring 2002

Introduction

The end of the second millennium was merely a numerical occasion that left no significant traces on humanity. Even the date conversion of computer times was taken in stride by our worldwide information and communication networks. Instead, the widely anticipated catastrophes and dislocations took place at other times; September 11, 2001, was a far more consequential event than was January 1, 2000. This upholds the thesis expounded in this book: Apart from the external technical clock time, there exists an intrinsic systems time that controls the processes of structural change, growth, and aging in systems ranging from organisms, populations, and institutions to states, cultures, and societies. Even the Universe is subject to phases of renewal and aging that are distinct from the external measured time. We therefore ought to spend less time watching the clock, and should take note of the inner temporal rhythms of nature and society.

What becomes apparent is a complex network of temporal rhythms in which physical, biological, psychological, and social processes superimpose and interact with each other. According to our present understanding, time is an interdisciplinary concept *par excellence,* in which the natural and social sciences and the humanities complement and rely on each other. This is the novel perspective that lends this book its particular character.

How did the story of time begin? When did people first take an interest in time? The emergence of an awareness of time in hominids and the temporal orientation of early hunting and farming societies were followed by the first astronomical time determinations by the ancient high cultures. Pre-Socratic natural philosophers like Parmenides and Heraclitus were the first to formulate the fundamental questions that influence discussions of time to this day. Is the world, as Heraclitus believed, in a continuous state of creation with time running irreversibly like the current of a river, or is every change merely an illusion, as Parmenides argued, and time a reversible parameter of an immutable world of eternal natural laws? That time can have a beginning was an assumption made by Saint Augustine, who related it to a divine act of creation.

The physical measurement of time in the Universe was examined by Newton, who assumed that all clocks in the Universe could be synchronized to an absolute world-time. Time measurements on Earth seemed to confirm this assumption. As to the direction of time, classical physics explained for the first time the fundamental time symmetry of physical laws, which permits events to proceed from future to past as well as from past to future. Youth and old age appear to be of no concern to physics.

In Einstein's theory of relativity, the measurement of time is no longer absolute but becomes a function of motion; the speed of light now represents a constant of nature, as well as the upper limit of velocity. A new quantity called *proper time* ("*eigenzeit*") is used to relate the different subjective times in systems moving relative to each other. Relativity theory uncovers startling effects: according to the *twin paradox*, persons that move at different speeds age at different rates. But because of the symmetry of time, the twins might just as well get younger at different rates! For Einstein, the direction of time was just a fantasy. His world is the modern version of Parmenides's world of immutable natural laws. But as Roger Penrose and Stephen Hawking have shown, Einstein's general theory of relativity also leads us to accept an expanding Universe with a singularity as its starting point. Time as an act of creation, after all, as claimed by Saint Augustine? This might appeal to theologians; not, however, to mathematicians and physicists (although this has nothing to do with their religious convictions). Instead, singularities represent a capitulation of mathematics: at singularities of vanishing size and infinite potential, the laws of physics can be shown to fail.

On subatomic scales, the laws of quantum mechanics and their dependence on Planck's constant must also be taken into account. It is remarkable that the laws of quantum mechanics are also symmetrical in time. To be sure, a brief violation of time symmetry may occur in the decay of a particular elementary particle if the so-called *PCT theorem* of quantum mechanics is indeed universally valid. However, a unified theory of quantum mechanics and general relativity is still lacking. Will the recently proposed *superstring theory* shed light on the mystery of the beginning of time? Until then, Hawking's assumption of an elemental Universe in imaginary time remains a mathematically intelligent speculation.

Detailed observations have now firmly established that the cosmic expansion proceeded from a virtually homogeneous initial state of strings and elementary particles to increasingly diverse and complex galactic structures. Does this represent the cosmic time arrow? And where does it point? Measurements of microwave background radiation force us to conclude that we live in a Universe expanding into a void. In the course of that expansion, very massive stars collapse into black holes. These lose energy and mass through radiation and will eventually disappear, taking with them the history of their stars. Once this has been physically confirmed, we will be faced with another problem, for in their places gaps in memory remain in the Universe. Is the expanding Universe, then, in a process of aging and afflicted with a "cosmic Alzheimer's disease," or do quantum fluctuations offer unforeseen rejuvenating cures or even an escape route to universes less inimical to life? Lacking additional observational data, the laws of quantum mechanics permit all kinds of time travel into a variety of conceivable science fiction worlds.

The assumption of a cosmic time arrow is at least in harmony with the second law of thermodynamics, in which physics first came to grips with the irreversible processes of nature: a cup falls on the floor and shatters, milk is poured into coffee and light coffee results, a star radiates light, etc. The reversals of these processes have never been observed. From a natural philosophy perspective, this is reminiscent of the Heraclitian world. Ludwig Boltzmann's insights led to the statistical interpretation of the second law, according to which it is highly likely that order will develop into disorder, while the reverse is unlikely. (A glance at my desk seems to confirm this trend.) To be sure, the validity of the second law is confined to systems in thermal equilibrium that do not interact with their environment; it cannot explain, for example, how weather or climate develop as conditions on Earth change. That requires nonequilibrium thermodynamics, which deals with the temporal development of open dissipative systems. While such systems may stabilize temporarily in local equilibrium states and create ordered structures, they may in the next instant and under changing conditions collapse or transform into new equilibrium states with different ordered structures. The dynamical properties of self-organizing systems like those known from chemistry and biology exemplify irreversible processes that, under changing conditions, may wind up at different attractors, ranging from fixed points and oscillations all the way to chaos.

Charles Darwin and Herbert Spencer's theory of evolution was the first to relate growth and life to increasing complexity. The evolution of life is based on irreversible developments in which particular plant and animal species may stabilize temporarily in the local equilibrium states offered by ecological niches. Genetic changes (mutations) permit new developments, which under the influence of selection pressures may either stabilize again, or else be discontinued. As conditions changed in the course of the Earth's history, very complex organisms have come into existence, while others have died out. Entire populations come to life, mature, and die, and in this they are like individual organisms. But while the sequence of generations surely represents the quintessential time arrow of life, many other distinct biological time rhythms are discernable. These rhythms are superimposed in complex hierarchies of time scales. They include the temporal rhythms of individual organisms, ranging from biochemical reaction times to heartbeats to jet lag, as well as the geological and cosmic rhythms of ecosystems.

In the physical laws that turn up in the natural sciences, time is just a parameter t representing a real, measurable quantity. But these laws remain valid when t is replaced by $-t$; that's what we mean when we say that natural laws are invariant under time reversal. How, then, are we to comprehend growth and decay processes that appear to have a direction in time? This question brings us to the core of this book's thesis. Complex systems that consist of many interacting elements, such as gases and liquids, or organisms and populations, may exhibit separate temporal developments in each of their numerous component systems. The complete state of a complex system is therefore determined by statistical distribution functions of many individual states. It has been proposed that time be defined as an operator which describes changes in the complete states of complex systems. This time operator would then represent the average age of the different component systems, each in its distinct stage of development. Accordingly, a 50-year-old could have the heart of a 40-year-old, but, as a smoker, have the lungs of a 90-year-old. Organs, arteries, bones, and muscles are in distinct states, each according to its particular condition and genetic predisposition. The time operator is thus intended to indicate the inner, or intrinsic, time of a complex system, not the external clock time.

The human brain may also be regarded as a complex system in which many neurons and different regions of the brain interact chemically and are switched among their component states. Our individual experience of "duration" and "change" thus reflects the complex-system states of the brain, which are themselves dependent on different sensory stimuli, emotional states, memories, and physiological processes. Hence, our subjective awareness of time is not contrary to the laws of science, but is a result of the dynamics of a complex system. This in no way diminishes the intimate appeal of a subjectively experienced flow of time as described in literature and poetry. Knowing the dynamical laws of the brain does not turn one into a Shakespeare or a Mozart. In this sense, the natural sciences and the humanities remain complementary.

The theory of complex systems also applies to the dynamics of socio-economic systems. A city, for example, is a complex residential region in which different districts and buildings have distinct traditions and histories. New York, Brasilia, and Rome are the result of distinct temporal development processes, which are not elucidated by external dates. The time operator of a city refers literally to the average age of many distinct stages of development. Institutions, states, and cultures are similarly subject to growth and aging processes, which external dates can shed little light on. From the point of view of complex dynamic systems, the discussion of age is not just metaphorical, but offers an explanation in terms of structural dynamics.

While the rise and fall of population groups and nations formerly remained local events, in the age of globalization they raise questions of the survival of humanity. The globalization of markets and the rise of the technological civilization mean that firms, nations, and continents now compete across national borders. Globalization's foundation is the computer time of the worldwide information and communication networks, which melds technical devices with humans and their brains, as if in a new superorganism. But computer time does not indicate the age of our species. That calls for social and cultural sensibilities, of which computer time is utterly ignorant.

While the fourth German edition of *The Little Book of Time* is released in Munich, this English translation is being published in New York. There is presumably no city better suited for capturing the international "pulse of time."

Time in the Classical and Medieval Worldviews

Over the course of their evolution, humans have developed an awareness of time so they could put in order the past, the present, and the future. Historically, the awareness of time was by no means universal or uniform. Ancient cultures all employed different models and techniques for measuring the duration and passage of time. In this chapter, we will briefly chart the intellectual history of time, exploring the early ideas that laid the groundwork upon which our contemporary concepts of time are based.

From the Beginnings to the Pre-Socratic School

The origins of human awareness of time are lost in the prehistory of primates. However, it is unlikely that we, *Homo sapiens sapiens*, were the only one of our species to develop an awareness of time. Over the course of 4 million years, the concept probably developed independently and differently in the various hominid species and subspecies, depending on the physiological evolution of each group's brain and its capacity for a long-term memory and abstract thought.

Some of the earliest evidence of human fascination with time is found in the Middle East. The ancient cultures on the Euphrates and Tigris employed astronomy to determine time, but their central astronomical interest was lunar observation, which formed the basis of a lunar calendar and a Moon cult. To this end, the Babylonians created precise tables of the heliacal moon risings (moonrise in near conjunction with the sunrise), which required taking into account complicated corrections related to the Moon's visibility, including the Sun–Moon distance, the Moon's periodic deviations from the ecliptic (the apparent pathway of the sun through the sky), and the seasonal variation of the inclination, or tilt, of the ecliptic in relation to the horizon. These charts allowed the Babylonians to predict those lunar eclipses that take place at full moon when the Moon is precisely in the ecliptic. Today, we know that the lunar eclipse occurs because the Earth casts its shadow on the Moon, a fact that was unknown to the Babylonians because they lacked a planetary model of the Solar System. Because the Babylonian calendar was lunar-based, it frequently fell out of step with the seasons, which follow the solar calendar. If the Babylonian king felt that the calendar and the seasons were too much out of sync, he would simply add an extra month to the calendar to correct the matter. Such were the powers of the king.

While we do not use the original Babylonian calendar today, of course, our division of the day into 24 hours, or $24 \times 60 = 1440$ minutes, and $24 \times 60 \times 60 = 86,400$ seconds, comes from Babylonia. The Babylonians employed a positional notation that is similar to the modern decimal system, but their notation was sexagesimal, i. e., based on powers of 60 instead of 10. Remnants of this system are still in use to this day: the circle is divided into 360 degrees, a degree into 60 minutes, and a minute into 60 seconds.[1]

In ancient Egypt, too, astronomy provided the means to mark time, although it was more the astronomy of the sun and stars than of the moon. The rising and setting of certain stars, owing to the daily rotation of the celestial sphere, were used to divide the night very roughly into 12 "hours." (The resultant units of time measured out by these so-called *star-clocks* do not of course correspond to the hour unit we use today.) The Egyptian year was also based on astronomy: it was divided into 36 decadic (10-day) periods each. A star's time of rising shifted by 1 "hour" every 10 days, and the star disappeared altogether after 120 days, only to reappear later.

While we know that the regular rising and setting of the stars is caused by the rotation of the Earth about its own axis and its revolutions around the Sun, the ancient Egyptians explained this movement through mythology. They believed that the sky reenacted the unchanging drama of the death and rebirth of star gods, including Osiris (Orion) and Sothis (Sirius). The heliacal rising of Sirius (i. e., just before sunrise) and the Nile flood, which marked the start of Egypt's fertile season, were linked to each other in the Egyptian sun calendar. The ancient cult of Isis, the goddess of love, related earthly fertility and life to the star mythology of resurrection and rebirth.

Our modern calendar is based on one of the developments of the ancient Egyptians. That calendar was known as the *Sothic period*, because the Egyptians determined it from the heliacal rising of Sothis (Sirius). The tropical solar year (or the mean solar year) from solstice to solstice lasts 365.2422 days and goes from noon to noon. Since a calendar only considers whole days, the Egyptians made the year 365 days long with 12 months of 3 decadic periods each, and left 5 extra days at the end. After 4 years, the solstices and the seasons had shifted by approximately 1 day. Over the course of 1460 years, they were pushed forward by an entire year.

Around 450 B.C., the lunar calendar inherited from the Sumerians (predecessors of the Babylonians) was combined with the solar calendar developed by the Egyptians, and for that purpose a "leap month" was added to the year's usual 12 months in 7 out of 19 years. This 19-year cycle, known today by its Greek name, the *Metonic cycle*, is of Babylonian origin. The cycle has a period of 235 months ($19 \times 12 + 7 = 235$), resulting in a year of 12.37 months ($235/19 = 12.37$). Consequently, the average month has not 30, but 29.1 days, compared to the 29.53 days that is today's value for the length of a *synodic* month, which lasts from full moon to full moon.

It may have been the observation of periodic celestial events that first gave humans the notion of the regularity of nature. This concept is connected with the idea of an unchanging unit for measuring time, to which the duration and variation of natural terrestrial events could be compared. This type of observation, and this sort of knowledge, must have opened the doors to highly abstract considerations of the nature of the world. The pre-Socratic natural philosophers were the first to inquire into the causes of change and of regularity, and it is with them that the philosophical and scientific contemplation of the concept of time began.

Heraclitus (c. 550–480 B.C.) considered change itself to be the basic stuff from which everything is made. According to his universal law (*logos*), the struggle of opposites tends toward harmonious fusion; Heraclitus's vision of a constantly changing river continuously carrying new water became famous. This has led several modern authors to credit Heraclitus with the discovery of irreversible, unrepeatable processes and, therefore, the *time arrow*. In doing so, however, they often overlook the fact that, according to Heraclitus, the law of change is itself unchanging and eternal.[2]

For Parmenides of Elea (c. 515–445 B.C.), reality was only what exists now, enduring and unchanging, perpetual and infinite, and not what was in the past or may come to be in the future. Change, he believed, is an illusion of the senses, and only permanence is real. Parmenides visualized this permanent existence as an ideal sphere that is immovable and uniform throughout. In the history of philosophy, the Eleatic doctrine of unvarying existence has been seen as criticism of Heraclitus's concept of constant change. To the Eleatic philosophers, change is not just ephemeral—it is nonexistent, a figment of the senses.

Zeno's Time Arrow and Aristotle's Continuum

Zeno of Elea (c. 490–430 B.C.), a student of Parmenides, defended the Eleatic doctrine of permanent existence by employing four paradoxes of change, which have been variously understood and interpreted in the course of the history of philosophy and science. Here we will focus on two of them.

Zeno's famous second paradox states that Achilles, the fastest athlete in the ancient world, can never overtake a tortoise. The paradox asserts that Achilles would give the tortoise a 10-meter head start because it was slower, but he would never be able to pass the animal. He must first reach the spot where the turtle started the race, and during that time, the tortoise would have gone a certain distance ahead. By the time Achilles reaches that spot, the tortoise has again gone ahead, so that Achilles keeps getting closer and closer to the creature, but can never overtake it.

This argument presupposes that any length must contain an infinite number of points, so that Achilles can overtake the tortoise only after an infinite number of instances. If the length is considered to be a true continuum, the conclusion that it is infinitely divisible is mathematically correct. But it would be mathematically incorrect to conclude that an infinite number of instants always adds up to an infinitely long time. An "infinite sum" (series) consisting of terms that get successively smaller may well converge to a finite magnitude:

$$\sum_{\nu=1}^{\infty} \frac{1}{2^\nu} = \frac{1}{2} + \frac{1}{4} + \frac{1}{8} + \ldots = 1$$

This mathematical objection was countered by Zeno and the Eleatics with the assertion that what mattered was not the mathematical paradoxes of the infinite, but the infinite divisibility of "real" time intervals and distances. However, modern physics does represent an interval by a continuum of numbers, so that the mathematical objection remains a valid one. The situation changes only if a minimum length is presupposed, as was suggested, for example, by Werner Heisenberg in his studies of quantum physics. Zeno's idea was to push *ad absurdum* the assumption of a length being indefinitely divisible, in order to reveal as illusion the assumption of a multitude of points. By doing so, Zeno intended to prove the Eleatic dogma of an indivisible existence.[3]

The Eleatic School dealt specifically with what they saw as the illusion of temporal change in the third paradox, which considers an arrow in flight: During an instant of time, an arrow in flight occupies a space in which it remains for the duration of this instant. In the next instant, too, the arrow occupies a space from which it does not move. How, then, can it ever move, no matter how short the interval between the two instants?

Again, it is easy to raise mathematical objections to Zeno's paradox if the flight distance of the arrow is considered as a continuum of numbers (as it is in modern physics). For it is impossible to speak of a "next" position of the arrow, because the points in the continuum are dense, i. e., between any two neighboring points one can always insert another point (e. g., by halving the distance between them). Against this, too, the philosophical objection was raised that Zeno was only concerned with "real" time, according to which we always live in the present (a Now), so that the paradox continues to stand.

In the history of philosophy, the atomic theory of Democritus is often presented as a consequence of the Heraclitian doctrine of change and the Parmenidean principle of an unchanging existence. The Parmenidean distinction between the "existent" and the "nonexistent" corresponds to the Democritian distinction between the "full" and the "empty," between the smallest indestructible atoms and empty space. Heraclitian diversity and change are explained as the rearrangements of atoms. While the objects we perceive, such as stones, plants, and animals, are macroscopic aggregates of atoms that can change in time and can combine into new atomic groupings, atoms and empty space are, according to Democritus, timeless, uncreated, and eternal.

The disciples of Pythagoras (c. 570/560–480 B.C.) also searched for the timeless and the unchanging in nature. They espoused the profound idea that it is not some set of basic materials that are timeless and eternal, but the mathematical laws governing them. The perfect geometrical forms were for Plato (427–347 B.C.) examples of the timeless fundamental images and ideas of things that our senses perceive as merely imperfect replicas. Just as a column of a Greek temple is only an imperfect replica of the geometrical ideal of a cylinder, an honest politician or a courageous soldier are merely imperfect replicas of the idea of justice and courage. In the *Timaeus* dialogue, Plato justified his doctrine of time and timelessness in the same way. The stars and sky that we perceive are a moving (in fact, rotating) image of eternity, a world of perfection that endures timelessly. According to Plato, time is determined by the number of rotations of the spheres; days and nights, months and years correspond to the complete and partial rotations of the heavenly bodies. In that sense, time only came into existence with the cosmos and is an imperfect replica of the timeless.[4]

Aristotle (384–322 B.C.) condemned Plato's distinction between a temporal cosmos and a timeless and eternal world of ideas as fictitious and artificial. Aristotle maintained that it is the task of physics to explain the principles and the functions of diversity and variability in nature. Aristotle used the terms *form* and *matter*: form was the universal, which makes a particular entity what it is—for example, a stone, a plant, or an animal—and matter was what is governed by form. But form and matter did not exist as such; they were rather principles of nature that are derived by abstraction. That is why matter was represented as the possibility, or the potential, of being formed. Only by matter becoming formed was *reality* created. *Movement* was, in general, considered as *transformation*, as a transition from the possibility to reality, as an "actualization of the potential" (as it would be put in the Middle Ages). According to Aristotle, movement encompassed all goal-oriented processes in nature, such as a stone falling to the ground, the growth of a tree from seed to maturity, and the development of a human being from infant to adult.

Aristotle presented the following solution to Zeno's paradox of Achilles and the tortoise: The distance they cover is only *potentially* infinitely divisible, but does not necessarily consist of an infinite number of real (actual) subdivisions. While the distance is accordingly divisible into many parts in a single continuous process, it is in fact not so divided, and as long as it is not, it is possible to run through it. For Aristotle, distance was a continuum (i. e., something that is "continuously connected"), in which an infinite number of cuts could potentially be made, but it did not actually consist of an infinite number of pieces. This Aristotelian distinction between potential and actual infinity playes a central role in discussing the foundations of modern mathematics.

Aristotle also solved Zeno's paradox of the time arrow. He objected particularly to Zeno's definition of the present. According to Aristotle, the Now is as little a part of time as a point is a part of a distance. One must instead picture the present as a potential cut, not an actual cut, in the time continuum. The present is accordingly not the time point actually occupied by the arrow. Only potentially does the arrow remain still for an instant of time. In actuality, the arrow executes a continuous motion.

Aristotle was the first philosopher to formulate the concept of a continuum precisely. Time, he said, is permanently connected to the Now, but time is not allotted an independent existence. Only the movements of nature are real, and the Now of an instant is a cut in the continuum of motion. Since it is possible to make a potentially unlimited number of cuts in the continuum, action results in an innumerably large number of instants without ever being able to exhaust the continuum. "Time," said Aristotle, "is, however, not movement, but its innumerable aspect."[5] As a measure of time, Aristotle suggested circular motion, which other motions can then be compared to. The assumed spherical motion of the stars and the planets are a good example. Just as the circle is a basic shape in Euclidean geometry, so is uniform circular motion the foundation of classical and medieval astronomy.

Along with his philosophy of time and continuity, Aristotle developed the first logical theory of time modalities. *Real* is what is realized (i. e., is true) *now*, this very instant. *Possible* is what is realized in the present or at a future time. *Necessary* is what is realized at any *future* time. The modalities are characterized by their relations to the Now. While the logical system of the Stoics accepted the definitions of temporal modalities, the Megarians (a school of philosophy founded by Euclid of Megara) employed the modalities of the possible and the necessary without relating them to the Now. *Possible* is what is realized at *any* time. *Necessary* is what is realized at *each* time.

The Aristotelian analysis of the sentence "Tomorrow a sea battle will take place" is renowned.[6] It is true now that tomorrow a sea battle will or will not take place. Therefore, doesn't it follow that either it is already true now that a sea battle will take place tomorrow, or that it is already true now that no sea battle will take place tomorrow? In other words, does it not follow that one of the two statements (we don't yet know which one) is already true today?

In the symbolism of time logic, p signifies a state, such as "a sea battle takes place." The symbol Np stands for the state described by p taking place "tomorrow." As a measure of the time one might, for example, specify noon, when the sun is highest. The symbol $N\neg p$ means that the negation of the state described by p, namely $\neg p$, takes place tomorrow; i. e., p does not take place tomorrow. The symbol v stands for the logical "or" that connects two statements that describe two states. Expressing the preceding paragraph in symbols of logic, the statement $N(p \; v \; \neg p)$ denotes that it is true now that the state denoted by p exists tomorrow or does not exist. Doesn't it then follow, so goes the Aristotelian problem, that $Np \; v \; N\neg p$; i. e., isn't it already true now that the state denoted by p exists tomorrow, or that its negation $\neg p$ exists tomorrow?

Modern logicians like Georg Henrik von Wright have shown that the response to this question depends on the topology of the temporal development assigned to the world. If time is conceived linearly, the time intervals (e. g., days) succeed each other on a straight line. In Figure 1.1, each small circle stands for an overall state (i. e., a conjunc-

FIGURE 1.1 Linear time with uniquely determined states of past, present, and future.
(After Kienzle and Bertram, ed., Zustand und Ereignis *[Frankfurt: Suhrkamp, 1994], 176.)*

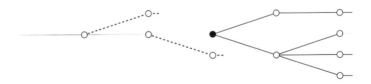

FIGURE 1.2 Tree of time with bifurcating possibilities of past, present, and future.
(After Kienzle and Bertram, ed., Zustand und Ereignis *[Frankfurt: Suhrkamp, 1994], 176.)*

tion of many partial states) of the world at a particular time. The filled circle ($\bullet$) stands for the overall state that is presently actualized. The circles ($\circ$) that follow represent the overall states in the future, and the preceding ones represent those in the past. In this model of time, the states that follow are determined uniquely and without alternatives, so that in this world it is already true that tomorrow a sea battle will take place, or it is already true that tomorrow no sea battle will take place.

If we admit the possibility of several development branches in the future, we obtain a time tree as illustrated in Figure 1.2, where the actualized Now ($\bullet$) is succeeded by several possible overall states ($\circ$) in the next time interval (e. g., a day), which may themselves be succeeded by several possible overall states in the following time interval, etc. The dashed lines indicate that several possible time developments may have taken place in the past, as well, but which were not realized. The past is, therefore, a linear sequence of overall states with respect to succeeding time intervals.

The statement that $N(p \vee \neg p)$ is true in a particular time interval is valid in both of these models of temporal development. For the present interval ($\bullet$), it is true that in both Figures 1.1 and 1.2, the partial state designated p is a component of the overall state in the following time interval or it is not ("*tertium non datur*"). The statement that $Np \vee N\neg p$ is true in a particular time interval is, however, generally valid only in a linear world (see Figure 1.1). But for the case of a branched future world (Figure 1.2), it would have to be true in the present time interval that the state p is a component of every overall state in the next time interval (Np). Alternatively, it would have to be true that the partial state p is not part of any of the overall states that follow ($N\neg p$).[7]

In the example of the ancient sea battle, the state of Np being already true is probably never fulfilled. But one can easily imagine conditions under which $N\neg p$ is true today already; for instance, all ships might be so widely distributed today that they could not possibly be assembled into a fleet tomorrow. If $N\neg p$ is true, it follows that $Np \lor N\neg p$. These considerations of modern temporal logic in the spirit of Aristotle show why he is rightfully considered one of the greatest occidental philosophers, whose thinking influences science to this day.

Time and Creation According to Saint Augustine

Not surprisingly, the Judeo-Christian tradition's concept of linear time is founded in religion. Time begins with God's creation of the world, and it ends with its demise on Judgment Day. God, the creator of the world, is Himself timeless, eternal, and uncreated. The Jewish Yahweh, on the other hand, is less a symbol of a mathematical world order and more the conception of a moral lawgiver.

The Judeo-Christian concept of time and the Platonic philosophy of time were combined in Neoplatonism. Plotinus (A.D. 205–270) was primarily concerned with the distinction between eternity and time, assigning eternity to the spirit of God and time to the realm of becoming—our Universe. Following Plato's example, the eternity of God was understood as the archetype whose turbulent replica is time. Thus God's spirit became the world's soul, which is manifested in the time of the turbulent Universe.

Saint Augustine's (A.D. 354–430) philosophy of time is accorded considerable significance in the Christian tradition. Plato's Demiurge, who had created the cosmos in mathematical harmony, became the Christian revelation's personified creator-god with whom Saint Augustine held dialogues in his *Confessions.* According to the *Confessions,* time was created with the cosmos, and Saint Augustine was critical of ideas in which time was identified with the spherical motions of the Universe, such as the orbital motion of the Sun around the Earth in the geocentric worldview. Even the rotations of the heavenly bodies are merely examples of motion, during which time elapses. According to Saint Augustine (and contrary to Aristotle), there can be no motion without time.

He argued if all the lights in the sky were to stand still, but a potter's wheel continues to turn, would time no longer exist? True, we can relate temporal processes to various movements of bodies, such as the Sun's orbit (or the Moon's orbit for the Babylonians); but in the final analysis, one must presuppose the time of the world. But what does it mean to measure time itself? Saint Augustine's famous answer was "Let us see then, thou soul of man, whether present time can be long: for to thee it is given to feel and to measure length of time."[8]

From the vantage point of the modern world, this citation has often been interpreted as the beginning of an individual's conception of time that is internalized, subjective, and even psychological, in contrast to the time of the external physical world. But as a matter of fact, such modern distinctions played no role in Neoplatonism. In the Plotinian tradition, the "soul" referred rather to the world's soul, which guides all movements of the cosmos. The measure of time of the cosmos was founded on the world's soul and not in the individual's subjective sense of time.[9]

Christianity, Judaism, and Islam belong to the same group of world religions as far as their concept of time is concerned. Time has a beginning and an end, running from the creation to the end of the world, and, in that interval, the variously interpreted story of salvation is enacted. Contrary to this linear conception, other religions advocate cyclic concepts of time. Hinduism proclaims the belief that souls migrate in the cyclical course of time, while Buddhism teaches how to overcome the cycle through self-denial by the Eightfold Way. The soul can then come to rest in the timeless state of Nirvana.

The Chinese tradition, on the contrary, is concerned not with surmounting time, but with dealing with time in a harmonious fashion. A timeless Beyond outside of nature and society is not a goal envisioned by the Chinese. The Taoist philosophy of nature teaches that people can live harmoniously within the great temporal rhythms of nature. Confucianism seeks ethical rules for a harmonious life within human society. As in the high cultures of the West, the measurement of time in China has depended on the state astronomical knowledge, and at least since the Middle Ages the Chinese have based their time measurements on spherical models for the motion of stars and planets in the sky.

Time and Medieval Astronomy

Among the Greek astronomers, the Pythagoreans were the first to suggest a geocentric model in which the Earth is orbited by the Sun, the Moon, and the five planets known at that time, whose different orbital periods were explained by their various distances from the Earth. The greatest mathematical precision in classical/medieval astronomy was attained by Ptolemy (c. A.D. 100–170), who explained the varying lengths of the seasons by the apparent acceleration or deceleration of the orbiting Sun. The Sun, understood as a planet, was said to move uniformly with respect to the imagined center of the celestial sphere of its orbit. To an observer on Earth, which was assumed to be located eccentrically with respect to the center of the sphere, the Sun *appears* to accelerate or decelerate, depending on its distance from the observer.[10] Even today, we relate astronomical data of position and time to a geocentric model. However, the Earth is not taken to be the center of the world, but merely the origin of the coordinate system we selected for determining positional and temporal data.

As did the ancient Chinese, we use the celestial equator (i. e., the projection of the terrestrial equator) as the base circle, together with its celestial north and south poles, and we use the vernal equinox as the directional reference point (origin). Greek astronomy, on the other hand, chose the ecliptic, i. e., the solar orbit, as the base circle with its north and south poles and with the spring equinox as the origin.

Periodic phenomena like the daily apparent revolution of the sky or of a celestial body like the Sun are obvious choices for units of time. The beginning and the end of a complete daily rotation are determined by the coincidence of a fixed mark on Earth and an agreed-upon direction in the sky. The local meridian is chosen as a fixed point, and one uses either the vernal equinox point or the center of the solar disk as reference point in the sky.

Since the Sun travels at a nonuniform speed along the ecliptic (which is tilted from the equator), it does not lend itself to time measurement without some adjustment. In order to eliminate the deviations caused by the apparent variable orbital speed of the Sun, time measurements today are based on a fictitious, or imaginary, *mean Sun*, which completes an orbit at the same average speed and in the same time (i. e., in one year) as the real Sun. A mean solar day is the average value of all of one year's solar days of unequal lengths. A true solar day is the time interval between two successive transits of the local meridian by the Sun (at midnight, below the horizon). A simple sundial indicates the true solar time, the so-called *hour angle* of the true Sun.

Calendars and Clocks

The basis of our modern calendar is the fixed solar year. Julius Caesar had, at the suggestion of Greek astronomers, improved the Egyptian solar calendar by adding a leap day every four years at the end of the Roman calendar (i. e., to February). Furthermore, one day was eliminated from February, so that the two months named after Julius Caesar and Augustus could have the same number of days, 31 each. That simple leap rule gave the Julian calendar year a length of 365.25 mean solar days. Since the solar year is shorter than the Julian calendar year by 0.0078 days, by the sixteenth century, an error of nearly 13 days at the beginning of the year had accumulated since Julius Caesar introduced the calendar in 46 BC.

This error was corrected under Pope Gregory XIII by having October 15 succeed October 4, 1582, without interrupting the normal sequence of the days of the week. The beginning of spring was defined to be March 21. According to the new leap rule, leap years are years whose last two digits are divisible by 4. To correct for the slightly shorter length of the solar year, 3 leap years are omitted every 400 years, and to that end, leap days are omitted in the secular years whose unit is not divisible by 4. Accordingly, the years 1700, 1800, and 1900 were not leap years, but 2000 was one again. It will take 3333 years before the remaining errors will have grown to a whole day.

Calendars are culture-dependent measures of time.[11] The numbering of years since the birth of Christ was introduced in the year 525 at the suggestion of abbot Dionysus Exiguus, while the Julian calendar had counted the years "*ab urbe condita*," i. e., since the founding of Rome. A central orientation of the Christian Middle Ages was provided by the calculation of feast days. Thus it had been decided as early as the First Council of Nicaea (A.D. 325) that Easter would be celebrated on the first Sunday after the full moon following the vernal equinox (i. e., when day and night are of equal length). In contrast to this, the religious life of Islam (in the Babylonian tradition) is determined by a lunar calendar. The Jewish calendar is based on a combination of lunar and solar calendars (the *lunisolar year*), which takes into account both the phases of the Moon and the passing of the seasons during the year. The insertion of a periodic thirteenth leap month ensures that the months remain in phase with the Moon, but the beginning of the year stays fixed except for small fluctuations.

To measure units of time smaller than a day, clocks are required. Ancient water clocks displayed time by use of a numbered, divided circle. Sundials indicate the cyclical motion of the Sun. An astronomical measurement instrument that is based directly on the celestial spheres of antiquity is the astrolabe, which was described, for example, by Ptolemy and by Theon of Alexandria. While it is a Greek invention, it blossomed under Islamic astronomers, who developed the mechanically most sophisticated specimens. The mechanical geared clock was invented between 1300 and 1350 but was only slowly accepted for use in everyday life, possibly because of its initial imprecision. The sandglass, which came into use at about the same time, was more familiar to the workman as well as to the seafarer. When held in Death's bony hand, it came to symbolize the transitory nature of life. For scholars, the mechanical clock was an object of great fascination from the start, for it almost seemed to be the realization of the Aristotelian definition of time: The steady rotary movement of the clock's hand is counted, stepwise, by a geared clockwork. Indeed, in 1377, Nicolas Oresme, the great natural philosopher of the late scholastic era, described the Universe as a geared clock whose escapement balances out all forces of the Universe. This marked the birth of the concept of a universal, mechanized time—a concept that modern science was never to let go of again.

1   Neugebauer, History of Ancient Mathematical Astronomy (New York: Springer-
    Verlag, 1976); Mainzer, Geschichte der Geometrie (Mannheim, Germany:
    Bibliographisches Institut & F. A. Brockhaus AG, 1980), 24.

2   Diels and Kranz, Die Fragmente der Vorsokratiker (Berlin, Germany:
    Weidmann, 10th ed. 1960–61), 22 B 64, B 30.

3   Ferber, Zenons Paradoxien und die Struktur von Raum und Zeit (Munich,
    Germany: C. H. Beck, 1981).

4   Plato, Timaeus 37 D,c, trans. Cornford, ed. Piest (Upper Saddle River, New
    Jersey: Prentice Hall, 1997).

5   Aristotle, "Physica," iv.219b, The Works of Aristotle I–XXIII, eds. Ross and
    Smith (London: Routledge, 1926).

6   Aristotle, "De interpretatione," 9 (19a28–32), Categories—On
    Interpretation—Prior Analytics, trans. Cooke, ed. Tredennik (Cambridge,
    Massachusetts: Harvard University Press, 1938), 11–179. Also compare
    Rescher and Urquhart, Temporal Logic (New York: Springer-Verlag, 1971),
    1–67.

7   Von Wright, Time, Change, and Contradiction: The Twenty-second Arthur
    Stanley Eddington Memorial Lecture, Delivered at Cambridge University, 1
    November 1968 (Cambridge, Massachusetts: Cambridge University Press,
    1968).

8   Saint Augustine, Confessions, Book XI, trans. Chadwick (New York: Oxford
    University Press, reprint ed. 1998), 11.15.19.

9   Flasch, Was ist Zeit? Augustinus von Hippo—Das XI. Buch der Confessiones:
    Historisch-Philosophische Studie (Frankfurt, Germany: Klostermann, 1993).

10 Mainzer and Mittelstraß, „Ptolemaios," Enzyklopädie Philosophie und
    Wissenschaftstheorie Vol. 3, ed. Mittelstraß (Stuttgart: Metzler, 1995).

11 Borst, "Computus: Zeit und Zahl im Mittelalter," Deutsches Archiv für die
    Erforschung des Mittelalters 44 (1988): 1–82.

Time in the Worldview of Classical Physics

In classical physics, time became a measurable and calculable quantity. Technical advances in mechanical engineering made possible the construction of increasingly precise chronometers and clocks. With the aid of modern mathematics, time could be measured with arbitrary precision. In the formalism of classical mechanics, time is merely a coordinate in the equations of motion. These equations remain valid under certain transformations—for example, those in which the direction of time is reversed. But the invariance of time as a constant measurable quantity—the notion that time is an absolute, independent, unvarying entity—was fundamental in classical mechanics. Only with the advent of relativity and then quantum theory were these notions of an invariant and independent entity called into question. They have also been considered by modern epistemology, where time is investigated as a form of consciousness.

Absolute Time According to Newton

The Copernican revolution, which replaced a geocentric Universe with a heliocentric one, is often seen as marking the beginning of the modern worldview. But Nicolaus Copernicus (1473–1543) was, in truth, very deeply committed to the ancient concepts of time and motion, according to which all planets moved at uniform speeds on spherical surfaces—meaning that the planets' orbits were circular. The classical and medieval faith in spheres would not be abandoned until Johannes Kepler (1571–1630) was able to provide better observational data. According to Kepler, the planets traveled in elliptical orbits and the Sun–planet radius swept out equal areas in equal times. In Kepler's laws, time is introduced as a measure of motion.

Galileo Galilei (1564–1642) was the first to define time and motion as basic concepts of modern mechanics. In modern mechanics, *velocity* is a quantity that is determined by the temporal change as a body moves from one position to another; *acceleration* is the change in a body's velocity in a given time interval. The concept of uniform acceleration, which Galileo initially assumed was for describing the motion of freely falling bodies (an idea that was later confirmed experimentally by means of the inclined plane), is now fundamental. Such temporal changes in velocity arise as a result of forces that affect the body in motion; for example, the gravitational force governs the acceleration of an object in free fall.

In principle, the time interval for successive positions can be made progressively shorter in order to let the velocity approach ever more closely its instantaneous value. In an arbitrarily short time interval, Δt, the change in position, Δx, also becomes arbitrarily small. The instantaneous velocity is then given mathematically by the quotient dx/dt, the infinitesimal change in position dx divided by the infinitesimal time interval dt. Reflections of this kind led Galileo, Kepler, and others to constitute the beginnings of calculus. Gottfried Wilhelm Leibniz and Sir Isaac Newton would later be the first to develop the mathematical rules for calculating with such "infinitesimal" quantities.[1]

In modern differential calculus, the instantaneous velocity is the limiting value of a sequence of average velocities for increasingly smaller time intervals, whose duration approaches 0. More succinctly, the instantaneous velocity is the first derivative of the position coordinate $x(t)$ with respect to time t. It is written in Newton's notation as $\dot{x}$ and in Leibniz's notation as the differential quotient dx/dt. The instantaneous acceleration is, analogously, the limiting value of a sequence of average accelerations for progressively shorter time intervals whose duration approaches 0. Because acceleration or change in velocity is a temporal change of a temporal change of a body's position, it is expressed as the second derivative of the position coordinate $x(t)$ and is written $\ddot{x}$ in conformity with Newton, and d^2x/dt^2 following Leibniz.

Galileo's contemplation of time measurement is noteworthy, as well. His suggestion for counting the number of swings of a clock's pendulum was later improved upon and made more precise by Dutch physicist Christiaan Huygens (1629–1695). Earlier, fourteenth-century French physicist and mathematician Nicolas Oresme introduced time as a coordinate and the mathematical representation of time intervals by geometrical distances. In the seventeenth century, Isaac Barrow, a teacher of Newton's, was the first to speak of time as a universal or "absolute" variable of nature. Barrow argued that time was independent of the different methods of observation and measurement and could be represented by a geometrical straight line. This concept led to Newton's famous distinction between absolute and relative time, with which he prefaced his treatise on mechanics, *Philosophiae Naturalis Principia Mathematica.*[2]

Newton also stated that a uniform motion, by which time can be measured precisely, may not exist because all motions are in fact accelerated or decelerated. Therefore, Newton introduced *absolute time*, as what in modern science would be called a *theoretical quantity*. This notion of time is well defined in classical mechanics, although it does not correspond to any direct experience. According to this assumption, all natural events can be related to a universal time whose topological structure (i. e., the chronological sequence of events) and metric structure (i. e., the units of time) are, in principle, constant. Newton also emphasized that it is a task of science, as it progresses, to continually improve the practical techniques for measuring time.

Under the influence of the Cambridge philosopher Henry More, Newton also connected his assumption of absolute space and absolute time with metaphysical interpretations of the omnipresence of God. But his concept of absolute time did not rely on this metaphysical component; its fundamental physical consequences were revealed in its precise mathematical formulation. What is of fundamental importance is the assumption that it is possible to determine objectively whether two events are simultaneous and whether they take place at the same location.

It follows from the Newtonian world's causality structure (i. e., from the connection between different events) that if two bullets are fired with different velocities from a particular temporal point O, they can only reach temporal points that are later than O. As a result, an event that takes place at O can only influence events that take place in the future. The past is no longer within one's sphere of influence.

This property of space-time—that an event's future and past have a common boundary, namely the present—expresses Newton's assumption that arbitrarily fast methods for transmitting time exist. One method for demonstrating the instantaneous transmission of time from position A to position B is to give a tug to a rod at A which is then transmitted to B without delay. The psychological reason for believing in this kind of simultaneity is that in everyday life, one's center of awareness naturally encompasses the things we see around us. That is, we extend our own time out to the entire world that enters our sphere of perception. A related assumption is the belief in a point that is at absolute rest.

Relational Time According to Leibniz

While the metrics and the corresponding causal structure of
Newtonian space-time were widely accepted in the eigh-
teenth and nineteenth centuries, Newton's assumptions of
absolute rest and motion were soon questioned. Theoretical
constructs like absolute time offered a good opportunity to
demonstrate inconsistencies in Newton's empirical approach to science ("*Hypotheses non fingo*"—"I do not make
up hypotheses.") and to accuse him of metaphysical speculation. For his contemporary Gottfried Wilhelm Leibniz
(1646–1716), space was a system of relations between
bodies. This system lacked any metaphysical or ontological
existence. The positional relationships between the bodies
were sufficient, according to Leibniz, to define space. In the
words of Newton, Leibniz dealt only with relative spaces or
reference systems.

Leibniz based the relativity of all space and time points
on his principle of sufficient reason—that is, the idea that
nothing happens in the world without sufficient reason.[3]
Mathematically, Leibniz's argument leads to a new spacetime symmetry, which differs from Newton's and in which
the concept of absolute rest and motion (rotation) must be
abandoned. Leibniz's space-time is less structured than
Newton's, but since the concept of simultaneity remains
unchanged, the causality structure of Leibniz's space-time is
the same as Newton's.

As long as one deals only with questions of kinematics
(i. e., questions that focus on the motion of bodies and not

the force that causes the motion), Leibniz's description of the world is exactly the same as the space-time of classical physics. Leibniz simply presented a kinematic relativity principle. But how does his principle explain dynamical effects, such as centrifugal forces in rotating systems? Leibniz and, particularly, his physics teacher Christiaan Huygens were aware of this problem. Huygens tried to explain the centrifugal force of a spinning disk by the relative motion of different elements of the disk. But the relative motion of the elements could be transformed away if one chose a reference system with the same origin and angular velocity as the disk itself, because relative to this rotating reference system, all elements of the disk are at rest. It is, however, well known that this fails to extinguish the pressures exerted by the centrifugal forces within the disk.

Time in Classical Mechanics

Leibniz and Newton were both correct in their mutual criticism. Newton's idea of a pole at absolute rest in the Universe cannot be validated by any observation or experiment, making Newton's space-time "too structured." But Leibniz's space-time has "too little structure" because the absolute rotational motions singled out by Newton definitely require a dynamical explanation. But does that make acceptance of the notion of absolute space necessary?

Leonhard Euler made the assumption that, without Newton's absolute space, it is impossible to formulate the law of inertia, which states that in the absence of external forces, a body moves in a straight line at a uniform speed. But following Ludwig Lange's introduction of *inertial systems* in 1885, the assumption of absolute space turned out to be superfluous. Relative to such inertial systems, Newton's law of inertia retains its physical meaning even without the assumption of an absolute space. For if one supposes that three mass points are ejected from the same origin point and are not subjected to external forces, then according to Lange's argument, the inertial system is defined by the three straight lines along which the mass points travel. Lange maintained that the law of inertia is then equivalent to the statement that relative to this system, any fourth mass point left to itself will also travel along a straight line. Today, one simply defines an inertial system as one in which Newton's law of inertia is valid. How such inertial systems (e. g., fundamental astronomical systems) may be identified by empirical methods is another issue.[4]

In contrast to absolute space, the time employed by classical mechanics applies uniformly to all inertial systems and is therefore an absolute time. In classical mechanics, all the points of an inertial system can be assigned the same synchronized measure of time. By the measure of time, $\widetilde{t}$, we understand only what is indicated by a clock that need not run at a constant rate, but its hands must never stop or go backwards. By the recalibration $\widetilde{t} \rightarrow t = t(\widetilde{t})$, we may then introduce a uniform measure of time t, defined except for additive and multiplicative constants, according to which all freely-moving point particles travel not only in straight lines, but also at a constant speed.

By a proper choice of the zero point and the units of time, finally, it can be arranged that clocks in different inertial systems are in agreement with each other. The assumption of absolute time in classical physics makes it possible to speak of a universal simultaneity that is independent of a particular inertial system, so that a particular point in time, $t = t_0$, separates the past from the future for all observers in the same way. The assumption of absolute time is expressed mathematically by the so-called *Galilean transformation* $x_i' = x_i(x_j, t)$ and $t = t'$, which connects the spatial coordinates x_j ($j = 1, 2, 3$) and the time coordinate t of an inertial system I with the corresponding coordinates x_i' and t' of an inertial system. Consider for example the transformation $x_i' = x_i + v_i t$, where the constants v_i represent the translational velocity of the inertial system I with respect to I'.[5] The form of an equation of motion in mechanics is therefore unchanged, or invariant, with respect to a Galilean transformation of all point coordinates. It is apparent that the Galilean invariance is more specialized than Leibniz's kinematic space-time because it confers preference to inertial motion. However, the Galilean invariance is more general than Newtonian space-time, since no system that is absolutely at rest is postulated.[6]

The time symmetry of classical mechanics is of fundamental importance. Newton's axiomatic equations of motion define acceleration as the second derivative with respect to time of the body's position. Expressed mathematically, time appears in the equation of motion raised to the second power, i. e., as t^2. It follows that if time, t, running in the forward direction is replaced by time running backward, $-t$ with a negative value, the equation of motion remains the same. In other words, the laws of mechanics are invariant with respect to the symmetry transformation $t \rightarrow -t$, and it follows that in classical mechanics, the two directions of time are indistinguishable. Every solution of an equation of motion with a positive time direction has a corresponding solution with a negative time direction.

One may visualize a sequence of events as a film that records the temporal development of the state of a system. Because of the time symmetry of classical physics, the laws of mechanics permit the film to run either backward or forward. Thus, according to the laws of planetary motion, a planet may orbit the Sun in the backward as well as the forward direction. On the basis of the time symmetry of the laws of mechanics, all mechanical processes without friction are reversible in principle.

But it is a fact that they proceed in only one direction. In an abundant number of processes, reversals have never been observed: a glass drops to the floor and shatters into many pieces; a tree grows from a seed into a mature tree; a person is born, grows older, and dies. The apparent irreversibility of such processes cannot be explained by the laws of mechanics. The time symmetry of mechanics seems closer to the static world of Parmenides, while irreversible processes are reminiscent of Heraclitus.

Various attempts have been made to explain irreversible processes on the basis of the improbability of their initial conditions. That the scattered shards should reassemble themselves into a glass is considered extremely unlikely, but is, at least in principle, possible. But the choice of (even improbable) initial conditions cannot explain why the process should proceed in this and not the other possible direction. The problem of the arrow of time, i. e., the breaking of the symmetry of time, is more fundamental, as will be seen later, in Chapters 5 and 6.

Time in Kant's Epistemology

Sir Isaac Newton's assumption of the existence of an absolute time, which does not correspond to any perception or direct measurement, produced various echoes in epistemology until well into the nineteenth century. In Immanuel Kant's (1724–1804) philosophy, this assumption was partially responsible for the idea of time being regarded not as an empirical reality, but instead as a form of our consciousness that precedes any experience ("*a priori*"), and which we must presuppose before we can observe, measure, or formulate physical laws. Human cognition, according to Kant, resulted from cooperation between sensibility and reason. By "sensibility," Kant meant our apparatus for perception and intuition, which locate the objects of our experience in space and time. In this view, our sensory organs merely deliver the material stimuli and signals of perception (e. g., light, color, sounds, pressure), which our intuition then orders according to spatial proximity and chronological sequence. Space and time, in this sense, are forms of our intuition which we use to organize sensory material. Concrete empirical clocks presuppose, according to Kant, the existence of time as an *a priori* form of intuition. On the other hand, we must not confuse time as a form of intuition with our subjective perception of time, which can vary in different situations and from person to person. For Kant, time for the general ordering of sequential events was an objective ("transcendental") form of intuition, which made possible the actual perception of time or the construction of empirical clocks.[7]

Aside from forms of intuition, we must also distinguish between conceptual and judgmental forms of reason. In Kant's epistemology, knowledge is gained by categorizing individual perceptions and intuitions ("manifestations") according to general concepts and thereby reaching judgments. For example, a manifestation or an image of the number 1 is the symbol /, which is depicted in various ways in different notations. Thus, the number 2 may be denoted by the symbol //, and hence quite generally the numerals /, //, ///, . . . may be assigned intuitively to the concept of natural numbers as images. The general scheme is then to add the unit / at each successive numbering step. According to Kant, this represented the general scheme for an *a priori* determination of time.

We might, for example, assign to each / unit one swing of a pendulum as an empirical unit, and then count the number of swings of a pendulum clock. But even without such an empirical assignment, the pure counting scheme remains, so that Kant's idea of time is a pure form of intuition and creates the *a priori* basis of arithmetic as the study of natural numbers. Mathematicians, including Sir William R. Hamilton (in the mid-nineteenth century) and Luitzen E. J. Brouwer (in the early twentieth century), have shared Kant's concept of time as the basis of arithmetic. The notion of time as a continuum, argued Kant, is based on our perceptions, which can grow continuously stronger or weaker with time. For example, we do not perceive the color blue as either blue or not blue, but, in a continuum of degrees, as more or less blue.

According to Kant, the modalities of time, such as persistence, chronological sequence, and simultaneity, determine which categories we use to judge our experiences. These categories include substance, causality, and interaction. From the *substance category* come the conservation laws, which recognize the existence of quantities that persist in time, or remain unchanged, such as energy. The *causality category* is the general form for physical laws of causality, which state that effects follow causes in time (e. g., the equations of motion). And finally, from the *interaction category*, come interaction laws, according to which it is possible for simultaneous events to interact. The actual formulation of the separate natural laws is the realm of physics, its experiments and its measurement techniques. Epistemology merely distinguishes, *a priori* and before any concrete experience, between the general forms of physical laws. Time was perceived by Kant as an *a priori* condition of categories, which must be presupposed for all observations, measurements, and formulations of laws and theories.

To be sure, it was already clear to Newton that the choice of particular units of time (days, hours, etc.) was arbitrary for two reasons. First, the choice is just a question of practical utility and not of real insight. Second, the choice is dependent on culture and technology. But Ernst Mach was the first to point out that we also cannot have an empirical intuition of the uniformity of a process. The uniformity of time implies that we can establish the equality of successive time intervals, but how can it be determined, for instance, that two successive swings of a pendulum take identical lengths of time?

In principle, no method exists for comparing successive time intervals registered by a clock, for it is impossible to transport the later interval back in time and place it alongside the earlier one. At best, one can place two clocks next to each other in order to observe if the beginning and end of their periods coincide. But it is not possible to determine by observation if two clocks will continue to run at the same rate forever. Nor can the laws of nature be used to determine if successive time periods are always of equal length, because time measurement is implicit in the validation of these laws. A unanimous chorus, from Ernst Mach to Jules Henri Poincaré to Hans Reichenbach, responded that the uniformity of time must be defined by means of a correlated standard (e.g., the Earth's rotation). The choice of a specific correlated standard (e. g., astronomical periods, atomic clocks) is again a question of practical utility and not one of true insight.[8]

In philosophy of science[9] an attempt was made to define a clock, without regard for its practical realization, as a device in which a point (the clock's hand) moves uniformly along a straight line. Before defining uniformity, the concept of *similarity* was defined, without reference to clocks, as a constant ratio of rates. The uniqueness of time measurement was then proved by the similarity of all clocks. Obviously, this definition of time is in the tradition of historical time measurement, as manufacturing techniques were just being developed for making clocks that ran at rates with constant ratios.

While the *uniformity* of time is concerned with the equality of successive time periods, *simultaneity* is concerned with the equality of two events that occur at the same time but at potentially different places. As we saw earlier, the simultaneity of two clocks at the same location can be tested, but how can the simultaneity of distant events be established? In essence, this requires a signaling link for transferring the time registered by a clock at one location to a second clock at another location.

As conceived by Newton, such a signal can be transmitted instantaneously, i. e., at infinitely high velocity. However, by the end of the nineteenth century it had been established that the fastest possible signaling speed is the finite velocity of light. Since the direct comparison of distant clocks is not possible, the following procedure for measuring the signaling velocity was suggested: At time t_1, a signal is sent from location A to location B, where it is reflected by a mirror at time t_2 and is received at A at time t_3 (see Figure 2.1).

The times t_1 and t_3 are registered by a clock at A, but all that is known *a priori* about time t_2 is that it is later than t_1 and earlier than t_3, i. e., that it is between t_1 and t_3. To calculate t_2, Albert Einstein suggested that it is the average of t_1 and t_3, i. e., $t_2 = t_1 + {}^1/_2 (t_3 - t_1)$.[10] This formula is based on the assumption that the signal takes the same time to travel from A to B as from B to A. But as Reichenbach pointed out, this assumption cannot be verified with the experimental arrangement of Figure 2.1. Furthermore, it raises the issue of defining simultaneity appropriately. From the finiteness of the signal velocity, it follows only that $t_2 = t_1 + \varepsilon (t_3 - t_1)$, with $0 < \varepsilon < 1$.

To enable measurements, correlating definitions of standards for the units, the uniformity, and the simultaneity of time are needed, and the requisite conventions should be chosen to make measurements as simple and practical as possible. The physical laws and theories are, of course, not affected by these conventions, since it should be possible to transform between different measurement techniques. But apart from the metrics of time, we must also consider the topology of time, i. e., the chronological sequence of time points. Here we must again distinguish between comparisons of time sequences at the same location and at different locations. With these epistemological considerations, we have arrived at the foundations of relativity theory, which revolutionized the physical concept of time.

FIGURE 2.1 A signal of light is sent at time t_1 from A to B, reflected at B at time t_2, and received at A at time t_3.

(After Reichenbach, Philosophie der Raum-Zeit-Lehre *[Berlin/Leipzig, Germany: Walter de Gruyter, 1928], Figure 18.)*

1   Ebbinghaus et al., *Numbers* (New York: Springer-Verlag, 1991), 307.

2   Newton, *Philosophiae Naturalis Principia Mathematica*, ed. Koyre (Cambridge, Massachusetts: Harvard University Press, 3rd ed. 1972).

3   Leibniz, *Hauptschriften zur Grundlegung der Philosophie* Vol.1, ed. Cassirer, transl. Buchenau (Leipzig, Germany: F. Meiner, 1904), 136.

4   Jammer, *Concepts of Space: The History of Theories of Space in Physics* (Mineola, New York: Dover Publications, 3rd ed. 1994).

5   Mittelstaedt, *Klassische Mechanik* (Heidelberg, Germany: Spektrum-Verlag, 1995), 47.

6   Audretsch and Mainzer, *Philosophie und Physik der Raum-Zeit* (Heidelberg, Germany: Spektrum-Verlag, 2nd ed. 1994), 28.

7   Kant, *Critique of Pure Reason*, eds. Guyer and Wood (Cambridge, Massachusetts: Cambridge University Press, 1999).

8   Reichenbach, *The Philosophy of Space and Time* Part 2, transl. M. Reichenbach (Mineola, New York: Dover Publications, 1982).

9   Janich, *Protophysics of Time: Constructive Foundation and History of Time Measurement* (Dordrecht, Netherlands: D. Reidel Publishing Company, 1985); Lorenzen, "Zur Definition der vier fundamentalen Meßgrößen," *Philosophia Naturalis* 16 (1976): 1–9.

10 Reichenbach, *The Philosophy of Space and Time* Part 2, transl. M. Reichenbach (Mineola, New York: Dover Publications, 1982).

chapter 3

Relativistic Spacetime

The concept of relativistic spacetime is fundamental to the structure of modern physics. The measurement of time is no longer absolute, but is, according to special relativity theory, path-dependent. In Einstein's terms, an observer at rest in one reference frame measures his own *proper time ("eigenzeit")* on a clock, while observers moving relative to him can calculate his proper time using their measurements of both distance and time. From Albert Einstein's general relativity theory—the relativistic theory of gravitation—it is possible to derive three standard models of the Universe that allow for both finite and infinite temporal developments, all starting from an initial singularity. In deciding between these models, quantum mechanics—the modern (nonclassical) theory of matter—must be taken into account. Newton's conception of time now remains valid only locally and approximately for velocities considerably lower than the velocity of light.

Time in Special Relativity Theory

The validity of Newton's equations in classical mechanics does not depend on any special inertial system, as long as the space and time coordinates are recalculated using Galilean transformations. In the spirit of Newton's research strategy, it was expected that Maxwell's electrodynamics equations would also prove to be Galileo-invariant. But after Heinrich Hertz, light had to be considered as an electromagnetic wave, and a simple calculation demonstrates that the wave equation of light is not Galileo-invariant.

Since we must accept the validity of the equations of electrodynamics on the basis of their overwhelming experimental confirmation (e. g., electromagnetism), we are left with two possibilities:

1 Mechanics is Galileo-invariant, but electrodynamics has a preferred frame of reference, in which ether, an element believed to fill the heavens, is located.

2 A principle of invariance (*relativity principle*) exists, which applies to both mechanics and electrodynamics. It cannot be the Galilean relativity principle. Hence, the laws of mechanics must be modified to include this principle.

Historically, Hendrik A. Lorentz and others supported the first alternative, while Einstein ultimately assumed the second. To validate the first alternative, it was necessary to establish the existence of ether, which moves relative to a reference system (our planet, for example). The renowned Michelson-Morley experiments of 1881 were designed to prove the validity of this assumption, but failed to obtain the sought-after effect, which would have established the ether's existence.

Unlike Lorentz, Einstein favored a single relativity principle common to both mechanics and electrodynamics. At the beginning of his famous 1905 paper on the electrodynamics of moving bodies, Einstein stated two postulates:[1]

1 Special relativity postulate: All inertial systems moving in straight lines and with constant velocity with respect to each other are physically equivalent.
2 Constant speed of light postulate: In (at least) one inertial system, the speed of light is constant, independent of the light source's state of motion.

The relativity postulate is a condition that had been formulated previously by physical theorists of the modern era (e. g., Huygens, Galileo). But while the constancy of the speed of light had not been confirmed experimentally at the time of Einstein's historic work, confirmations of the highest quality, based on experiments in elementary particle physics, are available today.

The spacetime structure that follows from Einstein's principles is determined by the appropriately defined transformations for inertial systems. In the absence of forces, the point particle moves, as before, linearly and uniformly with respect to an inertial system. The velocity of light must have the constant value c in every inertial system, in conformity with Maxwell's equations. Inertial systems characterized in this manner are called *Lorentz systems* and can no longer be related to each other by Galilean transformations. These are replaced by the *Lorentz transformations*, which leave invariant both the laws of electrodynamics and of mechanics (modified for high velocities).

Lorentz transformations no longer connect the time coordinates of different systems to each other as before. Thus, the time coordinate t' of a new reference frame I' is no longer just a function of the time coordinate t in reference frame I, but also of the space coordinates x_j ($j = 1, 2, 3$) of I; that is, $t' = t'(x_j, t)$. As a result, the concept of a universal time is lost, and with it the concept of universal simultaneity. Thus, it no longer makes sense to speak of space and time as absolutely distinct entities; one should, instead, envision four-dimensional spacetime as a single entity, the *Minkowski world*, named after Hermann Minkowski (1864–1909).

Whenever an observer establishes his Lorentz reference frame, he separates his spacetime into space and time as he measures them. Another observer, at rest with respect to another Lorentz frame, performs this separation differently, so that the observers in the two systems will disagree on their measurements of time and length. The time and the lengths measured in both systems can however be converted from one to the other by using the Lorentz transformations.

The action of bodies in a four-dimensional Lorentz frame, with Cartesian spatial coordinates x_j and the time coordinate t, may be visualized geometrically in the following way: If the speed of light is set equal to unity, $c = 1$, all particles traveling with the speed of light in a Lorentz frame travel in straight lines that make a 45-degree angle with the t-axis, and sweep out a cone whose equation, according to Pythagoras's theorem, is $t^2 = x_1{}^2 + x_2{}^2 + x_3{}^2$ (see Figure 3.1). Because of the constancy of the speed of light, all events in the future and the past must lie within this light cone and are considered to be *time-like events*. Particles with mass travel uniformly along straight lines as in Figure 3.1b, or non-uniformly along curves as in Figure 3.1c, always remaining within the light cone. Massless particles (photons) travel along the surface of the light cone, as indicated in Figure 3.1a.

In a Lorentz reference frame, the distance ("metric") from the origin O to a point Q with coordinates x_j and t is given by $(OQ)^2 = t^2 - x_1{}^2 - x_2{}^2 - x_3{}^2$ and therefore differs from the Euclidean metric for the same distance by the three minus signs. From the equation of the cone above it follows that if Q lies on the cone surface, $OQ = 0$, and if Q lies within the cone, then $OQ > 0$. Causal interactions between two events at two points P_1 and P_2 in the Minkowski world are restricted by the intersection of the future-cone of P_1 and the past-cone of P_2.

According to the new Lorentz time transformations, the measurement of time (by a clock) becomes path-dependent because of the clock's subsidiary dependence on the spatial

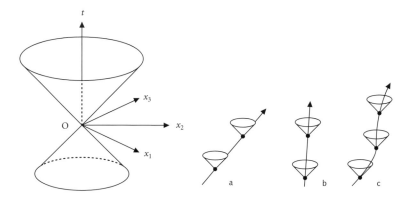

FIGURE 3.1 Future-cone and past-cone of an event at origin O in the Minkowski world for particles traveling at the speed of light (a), at slower uniform velocity (b), and at slower nonuniform velocity (c). All three spatial dimensions are represented in a plane orthogonal to the *t*-axis.

(*After Audretsch and Mainzer, eds.,* Philosophie und Physik der Raum-Zeit [*Mannheim, Germany: BI Wissenschaftsverlag, 2nd ed. 1994*], *39; Whitrow,* The Natural Philosophy of Time [*Oxford: Oxford University Press, 2nd ed. 1980*], *224.*)

coordinates. This consequence of the Lorentz transformation can be visualized by considering the age difference of twins who are in relative motion to one another: Two twins (initially the same age, as they've never done this before) are separated from each other. While the first remains stationary relative to a particular Lorentz frame, the other goes on a journey and, after some time, returns to his twin's location. At the moment of their reunion the first, stationary twin is older than his well-traveled twin. This effect has indeed been demonstrated in experiments using elementary particles traveling at relativistic speeds, which show clearly that everyone indeed has his or her proper time.

It must be emphasized that these laws of special relativity are once again invariant with respect to time reversal and are time-symmetrical, just like the laws of classical mechanics. As far as special relativity theory is concerned, it would be perfectly fine to envision the twins growing younger in the course of their separation, a possibility that we reject on the basis of our experience that all living beings do grow older. Special relativity theory is thus unable to explain the directionality of time and the irreversibility of the aging process.

It is worth noting that from the standpoint of Minkowski geometry, classical spacetime is not wrong. Einstein's theory can be restricted to a spacetime theory for inertial systems whose motion relative to the Newtonian inertial system of our Solar System is slow when compared to the speed of light. Inertial systems are, in any case, not too widely distributed in the Universe, and this restricted Einsteinian spacetime to be embedded into classical theory.

Time in General Relativity Theory

In order for special relativity theory to also accommodate the gravitational equation of Newtonian physics, it had to be expanded to the general theory of relativity. In 1907, when Einstein extended his investigation of spacetime to accelerated reference systems, he made the assumption that the acceleration effects in a reference frame are indistinguishable from the effects of a gravitational field. He illustrated this with his well-known thought experiment ("*Gedankenexperiment*") involving an observer inside a container who has no contact with the outside world. This observer can only observe the actions of bodies within his container. In a homogeneous gravitational field, all masses experience the same constant acceleration in a downward direction. Relative to a container traveling upward with the same acceleration, all bodies of arbitrary mass will experience the same downward acceleration. Therefore, an observer inside the container cannot determine from his measurements whether the container is being accelerated at a constant rate or is located in a homogeneous gravitational field.

One could also say that, as far as its effect on masses is concerned, a homogeneous gravitational field can be perfectly simulated by a suitably accelerated reference frame (*equivalence law*). Relative to a freely falling reference frame (e. g., an elevator), all effects of a homogeneous gravitational field can be eliminated, leaving the observer weightless.

So far we have only considered the special case of homogeneous gravitational fields. But an inhomogeneous gravitational field, in which the gravitational effects vary, can always be considered in regions that are sufficiently small for the gravitational effect to be almost constant, so that a homogeneous gravitational field is a good approximation. Therefore, according to the law of equivalence, the gravitational field is constant, at least in very small spacetime elements, and it is possible to select an inertial system that cancels the gravitational effects. The laws of special relativity without gravitation retain their validity locally. Einstein's *Gedankenexperiment* has today been performed by astronauts in orbit, who experience weightlessness during their free fall in the Earth's gravitational field.

The observer inside the container is also unable to use the measurement of a light beam's frequency as a means for distinguishing between a uniformly accelerating reference system and a homogeneous gravitational field. Just as a stone thrown into the air loses energy as it goes up, light loses energy when traveling through a gravitational field in a direction opposed to the gravitational attraction. As a result, its frequency decreases and its color is shifted towards the longer (redder) wavelengths of the spectrum. Similarly, if a light beam travels parallel to the floor of the container, from wall to opposite wall and at right angles to the direction of the acceleration or the gravitational attraction, it will be bent towards the container's floor to the same extent in both these cases. This is the same effect that Sir Arthur Stanley

Eddington observed when he measured the deflection of light beams from distant stars by the Sun's gravitational field.

In accordance with Einstein's equivalence principle, this example represents a case of local effects of gravitational fields that are inhomogeneous on a global scale. Freely falling bodies in inhomogeneous gravitational fields are subject to accelerations relative to each other. In the absence of gravity, free point particles and light beams travel with constant velocities and in straight lines, i. e., without experiencing accelerations with respect to each other. If a gravitational field is turned on by, say, bringing a large mass into the vicinity, their trajectories become curved, and the point particles are accelerated with respect to one another. Geometrically, the gravitational effect corresponds to curvilinear paths, and in that sense, a globally inhomogeneous gravitational field corresponds to a curved spacetime.

This is reminiscent of a well-known example from the spherical geometry of our terrestrial globe (i. e., geodesy). As is well known, the straightest path (and the shortest distance) between two locations on Earth is a circular arc on its surface. But if an arc is divided into short segments, each segment is approximately straight, so that we may consider the arc to be composed of infinitesimally short, straight segments. But the geometry of the straight segments is Euclidean, and for this reason, in Riemannian differential geometry of curved space (e. g., spherical surfaces), it is stated that locally, in infinitesimally small regions, Euclidean geometry is valid.

In an analogous fashion, the equivalence principle of Einstein's general relativity theory requires that locally, in infinitesimally small regions, the flat spacetime of Minkowski geometry applies, so that the physical laws are Lorentz-invariant, just as in special relativity. The physical laws are as a general rule covariant, i. e., they retain their form (form-invariance) in general coordinate transformation.[2]

The local Lorentz invariance determines the time and causality conditions in an overall relativistic gravitational field. If it is possible to connect a point P with a point P' by a time curve, then P can send a signal to P', but not the other way around. However, Einstein's gravitational equation also yields time-like solutions that are closed curves, and this kind of world would allow the physical paradox of an astronaut traveling into his own past. We will revisit this case in our discussion of relativistic cosmology.

Einstein's gravitational equation has been confirmed experimentally by measurements of the deflection of light around massive objects, transit time delay of radar signals, and the precession of Mercury's perihelion. In the context of this book, the predicted dilation of time due to gravitation is of particular interest. A clock located closer to the center of the Earth and hence deeper in its gravitational field runs measurably more slowly. This gravitational time dilation can again be visualized by a *Gedankenexperiment* involving the twins: A twin who had been exposed to a strong gravitational field on the surface of a very dense celestial body (e. g., a neutron star) will, upon his return to Earth, be appreciably younger than his twin who stayed on Earth. But neither general relativity theory nor special relativity gives preference to a direction of time ("growing older"), and a particular direction is assumed on the basis of so far unexplained experience.

Time in Relativistic Cosmology

Everyone on Earth can readily convince himself or herself of the symmetry of the Universe on a cosmic scale. For when we observe the starry sky, the naked eye, as well as the most powerful telescope, always perceives the same situation: an approximately uniform distribution of matter that is manifestly condensed into celestial bodies. Observations of this kind led to the general postulate of cosmology that, on average, matter is evenly distributed through the entire Universe (*homogeneity*), and that this phenomenon is independent of the direction of observation (*isotropy*). This homogeneity does not refer to the Universe in detail, but only to volume elements with dimensions of 10^8 to 10^9 light-years, within which matter may have condensed unevenly in the form of galaxies.

Has the Universe been uniform at all times, or does its symmetry only pertain to a particular period? In 1929, Edwin Hubble discovered that the velocities with which galaxies receded increased with their distance from the observer. He based this conclusion on the observation that light from very distant galaxies is shifted towards the red end of the spectrum, i. e., to longer wavelengths. Hubble interpreted this as an example of the *Doppler effect*, in which light emitted by a moving light source appears shifted to longer wavelengths if the source is receding from the observer, and to shorter, if it is approaching.

It follows from the cosmological principle that all points in space ought to experience the same physical development, correlated in time in such a way that all points at a certain distance from an observer appear to be at the same stage of development. In that sense, the spatial conditions in the Universe must appear to be homogeneous and isotropic to an observer at all times in the future and the past. Geometrically, this may be visualized by placing an observer at the center of the Milky Way who is equipped with a standard coordinate system with the directions of the spatial coordinate axes defined by, say, the lines of sight from the observer to particular galaxies. As the time coordinate one might select as a "cosmic clock" the radiation temperature of a black body, which decreases monotonically with time everywhere.

The supposition of the cosmological principle leads to Alexander Friedmann's standard models of cosmological evolution for the three geometric possibilities for homogeneous space, namely space with positive, zero (flat), or negative curvature.[3] Mathematically, the cosmic evolution according to these three standard models is described by a first-order differential equation derivable from Einstein's relativistic gravitational equation. The evolution is expressed in terms of a developing so-called *world radius $R(t)$*

and an energy density function. According to the singularity theorems of Roger Penrose (1965) and Stephen Hawking (1970), it follows from general relativity theory that the three cosmic standard models must include an initial space-time singularity with infinite curvature.[4] Cosmologically, this singularity is interpreted as the *Big Bang*, which causes the Universe to expand rapidly at first (*inflationary universe*) before slowing down. In the standard model with positive curvature, this expansion is eventually reversed and leads to a collapse, which represents a new singularity. This case is referred to as a *closed universe*. For the two other standard models, with flat and negative curvature, the expansion continues without limit and with lower and higher speeds, respectively, and these are referred to as *open universes*.[5]

At the time of the initial singularity, the density must have been infinite in the standard models. The 2.7-kelvin microwave background radiation, which is interpreted as a relic of this hot, early stage of the Universe, had been predicted as early as 1946 by George Gamow and was discovered by Robert Wilson and Arno Penzias in 1965. The microwave background radiation, whose spectrum has been shown to fit Max Planck's radiation formula for a black body precisely and whose intensity is almost isotropic, represents a compelling confirmation of Friedmann's three standard models.

In this context, it is important to appreciate the mathematical nature of a spacetime singularity like the Big Bang. Singularities have the drawback that physical laws are not defined in regions of spacetime with infinite curvature, so that no predictions about physical events are possible. Relativity theory leads directly to its own internal limitations. In the framework of Friedmann's models, it can only be asserted without clarification that time begins with the initial singularity. In the context of relativistic cosmology, time is merely a real coordinate for marking events, while the question of what occurred *before* the initial singularity is not defined mathematically and is therefore meaningless. Any talk of a "creation" of time is also mathematically undefined in the context of relativity theory. As a result, we must be careful to draw a sharp distinction between the defined concepts of a physical theory and how these concepts are interpreted in a particular worldview.

The singularity theories also predict the possible existence of very small regions of relativistic spacetime, in which spacetime becomes extremely curved and gravity becomes infinitely great. In astrophysical terms, such singularities are identified with *black holes*, which result from the death of a star by gravitational collapse. This means that there must exist a three-dimensional region in spacetime, bounded by the so-called *absolute event horizon*, which "swallows up" all incoming signals and from which no signals or particles can emerge. The spacetime singularity is assumed to be located at the center of the event horizon, which accordingly represents an absolute endpoint for causal time signals.

A *Gedankenexperiment* involving an astronaut traveling towards a black hole allows us to envision the consequences of Einstein's relativity theory. Suppose that an astronaut, on passing through the event horizon of the black hole, begins sending light signals to his space station outside the event horizon. If he sends the signals at constant intervals according to his clock, the space station will detect signals separated by increasingly longer intervals according to its clock, and the light will be increasingly red-shifted, until the astronaut's time, as seen from the outside, will come to a stop and his light signals will no longer be received because of the infinite curvature of spacetime.

General relativity, special relativity, and classical mechanics are all time-symmetrical theories. Their laws are unchanged when time is allowed to run backwards, i. e., when t is replaced by $-t$. Hence, general relativity theory also predicts the time-mirrored action of a black hole, that is, infinitely massive points that emit light signals explosively (*white holes*). But this mathematical consequence of a time-symmetrical theory is considered to be physically improbable, and Penrose excluded it by postulating an *ad hoc* hypothesis and exercising "cosmic censorship." The need for this hypothesis again demonstrates the internal limitations of relativistic cosmology.

Finally, among the other cosmological principles that have been proposed, the so-called *partial cosmological principle* of Kurt Gödel (1949) is of particular interest. According to this principle, the Universe is homogeneous, but not isotropic. The associated cosmic geometry admits the mathematical possibility of closed, time-like world lines that conjure up situations from first-rate science fiction. In Gödel's spacetime an observer could, for example, embark on a "journey into the past" in which he would encounter his earlier self. However, this suggestion of an anisotropic Universe is refuted experimentally by the well-documented isotropy of the microwave background radiation.

If the cosmological principle of Friedmann's models is indeed valid, the question arises how the initial singularity of time and the great symmetry of the Universe is to be explained physically. The cosmological principle and relativity theory are evidently inadequate for that purpose. In their place, modern cosmology now joins quantum mechanics and elementary particle physics in a research effort whose object is an explanation of the temporal evolution of the Universe.

1   Einstein, "Zur Elektrodynamik bewegter Körper," *Annalen der Physik* 17
    (1905), 891–921.

2   Weinberg, *Gravitation and Cosmology: Principles and Applications of the
    General Theory of Relativity* (New York: John Wiley & Sons, 1972), Chapter
    4.1: "The Principle of General Covariance."

3   Audretsch and Mainzer, *Vom Anfang der Welt* (Munich, Germany: C. H. Beck,
    2nd ed. 1990), 31.

4   Penrose, "Gravitational Collapse and Space-Time Singularities," *Physical
    Review Letters* 14 (1965), 57–9; Hawking and Penrose, "The Singularities of
    Gravitational Collapse and Cosmology," *Proceedings of the Royal Society of
    London* A 314 (1970), 529–48.

5   Audretsch and Mainzer, *Vom Anfang der Welt* (Munich, Germany: C. H. Beck,
    2nd ed. 1990), 93.

chapter 4

Time and the Quantum World

Although the measurement of physical magnitudes in the quantum world is limited in precision and is statistical in nature, time is still a parameter in a deterministic equation of motion. Called the *Schrödinger equation*, it is symmetrical in time, just like classical and relativistic mechanics. Consequently, it appears that the quantum world is also a kind of unchanging Parmenides world that lacks a preferred time direction. But in a single case, making a quantum mechanical measurement provides evidence of an irreversible process during which the temporal symmetry is broken. Possible violations of time symmetry also emerge in quantum field theories, which describe the interactions of elementary particles. The question arises, Will it ever be possible to explain irreversible processes within the framework of cosmic evolution, supposing that a union between general relativity theory and quantum mechanics can be achieved? It is suspected that there is an intimate connection between this epistemological discussion of time and many current research topics, including quantum mechanical measurement processes, black holes, and the anthropic principle.

Time in Quantum Mechanics

In 1900, Max Planck was obliged to introduce the minimum quantum of action named after him in order to avoid the infinity (*singularity*) that appears in the classical formula for the spectral distribution of a "black body" (e. g., a radiating oven). In 1905, Einstein employed Planck's assumption of a minimum quantum of energy in his explanation of the photoelectric effect. Later, Einstein also extended this theory to light quanta (photons), whose existence was confirmed experimentally in 1923 by the *Compton effect.* In 1913, Niels Bohr applied the quantum hypothesis to Rutherford's model of the atom in developing a theoretical explanation for the observed spectra of hydrogen and other atoms. To this end, he introduced the *correspondence principle* with which known results of classical mechanics—modified heuristically with Planck's quantum of action h—can be carried over into quantum theory. In this older quantum theory of Bohr, microphysical events take place in quantum jumps in which energy changes discontinuously by small, indivisible amounts (quanta).

In Hamiltonian classical mechanics, the state of a system is defined in terms of a pair of canonically conjugate measurable quantities, such as position and momentum. The temporal evolution of the system's state is uniquely determined by its Hamiltonian equations of motion, but in contrast to classical physics, quantum theory comprises rules that preclude an arbitrarily precise measurement of the state of the system.

The evolution of the states of an interactive quantum system (e. g., an atom or molecule) is uniquely determined by its time-dependent *Schrödinger equation*.[1] Since the Schrödinger equation is a partial differential equation, at first sight, the causality of a quantum system does not seem to differ mathematically from the causality of classical physics; this is also the reason that the Schrödinger equation is time-symmetrical. The difference between quantum mechanics and classical physics resides in the quantum states. In place of vectors, such as position and momentum, quantum theory employs operators for observable parameters. These operators obey certain exchange relations that depend on Planck's constant. The observable parameters in a quantum state can only be determined in terms of their statistical expectation values. In measuring two conjugate observables in a pair of systems that started in the same initial state, the results are distributed with standard deviations about their respective expectation values. Heisenberg's uncertainty relation states that the product of these two standard deviations cannot be smaller than a minimum value, which depends on Planck's constant h. For the deviations Δx and Δp of position variable x and momentum variable p, this is expressed as $\Delta x \, \Delta p \geq h/4\pi$.

In formal analogy to the uncertainty relationship for position and momentum, quantum mechanics also yields an uncertainty relationship for the (also canonically conjugate) variables of time and energy. Here, the variance in time is interpreted as the shortest measuring time. If two quantum states are to be distinguished by measuring their respective energies E, the uncertainty relation sets a lower limit for the requisite measuring time t, which depends on Planck's constant: $\Delta E \, \Delta t \geq h/4\pi$. The analogy to the uncertainty relation for position and momentum is, however, only a formal one, since no operator for time is defined in quantum mechanics. Time in quantum mechanics remains merely an invariant parameter, as in classical mechanics and in relativity theory, and is not a measurable quantity in the sense of a quantum mechanical operator. It is only later, when irreversible processes are considered, that it will be possible to define time, too, as an operator in the framework of a generalized quantum mechanics.

A decided departure from classical mechanics is presented by the *superposition principle*, which is an expression of the linearity of quantum dynamics. The superposition of two pure quantum states is, mathematically, a linear combination of the two, which again represents a pure quantum state. In Schrödinger's view, the two quantum states are superimposed or penetrate each other like two waves forming a wave packet (which once again represents a quantum state). The superposition principle is a linearity principle, and in that sense, quantum mechanics is a linear theory. Observables that still had definite values in the two discrete (pure) states of the quantum system possess only indefinite values in the superposition of the two states.

In the modern so-called *EPR experiments* (named after *E*instein, *P*odolsky, and *R*osen) it is possible, for example, to use polarized filters to analyze pairs of photons which a central source emits simultaneously and in opposite directions.[2] The correlations between their polarization states are considered to arise from the superposition of correlated photons. The two photons that had interacted with each other in the source remain in a strictly correlated common state even after leaving the source, even though they are now spatially separated and do not physically interact with each other.

Since the combined state of the photon pair can be calculated beforehand according to the superposition principle, it is sufficient to measure one photon's polarization at one filter in order to instantly predict the other photon's polarization at the other filter. In contrast to this, two tennis balls flying apart (e. g., after being shot from a tennis ball machine) are, according to classical mechanics, in separate states, which can be localized at all times. That is also why one refers to quantum mechanics as a "delocalized" theory, in contrast to "localized" classical mechanics.

The superposition (or linearity) principle of quantum mechanics has serious consequences for the temporal development of quantum systems. When the measurement starts, at time $t = 0$, both systems (i. e., the measured system and the measuring apparatus) are prepared in two separate states and their subsequent temporal development is determined by the Schrödinger equation. Because of the superposition principle, at a later time ($t > 0$) the complete state consists of nonseparable substates with indefinite eigenvalues. The measuring apparatus nevertheless yields a definite measured magnitude at time t. The linear time dynamics of quantum mechanics is thus unable to explain the measuring process.

According to the *Copenhagen interpretation of quantum mechanics*, the process of measurement is interpreted as the so-called *collapse of the wave function*; that is, the process of reading the measurement apparatus causes the superposition of the complete state ("superimposed wave functions") to split spontaneously into the separated states of the measurement apparatus and of the measured quantum system, thus yielding definite eigenvalues for both. But quite apart from the Copenhagen interpretation, it is necessary to draw a distinction between the linear temporal dynamics of quantum systems and the nonlinear act of measurement.

The Everett interpretation of quantum mechanics appears to avoid the problems of a nonlinear collapse of the wave function by connecting human consciousness to parallel but separate world developments (*many-world view*).[3] Everett argues that the state vector never splits into component states, but rather that all possible temporal development branches are realized in parallel. The complete state describes the multiplicity of concurrently existing real worlds, and every relative state of a world depends on the state of the measuring instrument, i. e., the state of the observer. To be sure, according to Everett's interpretation, the observer is only aware of a single temporal development branch, while the other parallel worlds are unobservable.

The advantage of Everett's interpretation is that the nonlinear collapse of superpositions need not be explained, but its obvious drawback is that it requires faith in the existence of a myriad of parallel temporal developments that are, in principle, unobservable.

The reason for such conflicting interpretations is that the Schrödinger equation, which describes the temporal development of the measured quantum system and of the measuring apparatus during the measurement, does not yield separate final states for the two component systems. Consequently, some physicists, in the tradition of Einstein, suspect that the linear dynamics of quantum mechanics is insufficient as a theory of matter. Their goal is a unified theory of linear quantum mechanics and nonlinear relativity capable of explaining the separate states of macroscopic systems without having to resort to human intervention, e. g., human consciousness.

According to a suggestion of Roger Penrose, the collapse of superpositions is caused by a significant amount of curvature of spacetime. The idea is that the rules of linear superpositions become modified when applied to gravitons as the smallest units of curvature, and some kind of time-asymmetrical nonlinear instability sets in. The one-graviton level of nonlinear collapses of superpositions lies between the quantum level of the elementary particles, atoms, molecules, etc. with linear dynamics, and the level of macroscopic everyday experience described by classical physics.[4] However, no verifiable unified theory corresponding to this idea exists to date. There are other suggestions for a general-

ized quantum mechanics, in which the superposition principle is restricted to make it possible to explain the nonlinear
separation of quantum systems on the one hand, and the
localization of macroscopic systems on the other. The measuring apparatus in the quantum mechanical measuring
process is thus conceived as a macroscopic dissipative
system.

No matter what kind of satisfactory explanation may one
day turn up, this much has already become apparent: While
(linear) quantum mechanics presupposes the time-
reversible and deterministic dynamics of quantum states,
measurements and observations are irreversible processes,
which incidentally select a time direction. For if after obtain-
ing an observational result, one attempts to determine the
former state of the system using the same method, the results
will be incorrect.

From the point of view of epistemology, it is interesting
to note that the quantum mechanical measurement process
is related to Zeno's philosophical paradox of the time arrow
described in Chapter 1. According to Zeno, the motion of an
arrow in flight is an illusion, for during a miniscule instant
(e. g., during an observation) the arrow is in a state of rest.
As we shorten the interval to the next instant, the observed
change in position decreases. Since the intervals to the next
instant can, in principle, be shortened at will, there will be
no change in the arrow's motion in the limit. In a 1977 paper
entitled "Zeno's Paradox in Quantum Mechanics," Baidya-
nath Misra and George Sudarshan described a process in
which an unstable atomic nucleus prone to radioactive

decay is prevented from decaying by continuous measurements and observation.[5]

In quantum measurements, the "quantum Zeno effect"
may be used to "freeze" a quantum system in its initial state.[6]
This has been demonstrated for an isolated special atom
with three specific energy levels. Instead of observing the
state of motion of Zeno's arrow, researchers investigated the
energy levels of the atom. A radio signal was used to excite
an electron from the atom's previously prepared ground
state to a metastable state, and the populations of the two
possible states was measured by a simple laser-based tech-
nique without affecting the populations themselves.

During the unobserved interval between successive
measurements, quantum mechanics cannot rigorously
predict which of the states is occupied, but yields only a
superposition of the two possibilities. (The temporal development of this superposition while the atom is exposed to
the radio frequency field corresponds to the flight of Zeno's
arrow.) If one attempts to measure the state populations
within that interval by employing additional laser pulses, the
probability of finding the electron in the metastable state at
the end of the interval drops sharply with the number of
additional measurements. Statistically, less than 1 electron
in 100 will be able to reach the metastable state if the number of observations is increased 64-fold.

According to the Copenhagen interpretation, every
measurement process annihilates the superposition and
forces the electron to select one of the two states. Following
each measurement, a new superposition is created until it is

annihilated by the next measurement. If these perturbations occur frequently enough, the electron is scarcely able to leave its initial state, in spite of being irradiated by the radio frequency field. As Zeno would say, the arrow is scarcely able to fly on: it remains at rest.

But any mention of a mysterious "collapse of superpositions" as a result of the measurement process is, as we have seen, an embarrassment for linear quantum mechanics. As a matter of fact, the quantum Zeno effect in the experiment described above can also be explained without that postulate. A measurement in that experiment is accordingly nothing more than an irreversible dynamic development of a quantum system under the influence of an optical laser field.

Time in Quantum Field Theories

The basic theme of quantum electrodynamics is the interaction of elementary particles or their respective matter waves with electromagnetic fields.[7] An electric field may be visualized as arising from electric charges. If the charges move, a magnetic field is generated whose distribution is described completely by the magnetic potential of the electric field. Every local variation of the electric potential can be compensated by a change in the magnetic potential, so that the electromagnetic field as a whole remains unchanged, i. e., is *invariant*.

We are already familiar with electromagnetic interactions from everyday life. Thus, the emission of electromagnetic waves caused by accelerated electrical charges in radio antennas and X-ray tubes are well known. The "weak interactions" within atoms, on the other hand, are observed far more rarely. For example, in the beta decay of the neutron, a neutron is transformed into a proton and an electron–antineutrino pair. At first, it seems that weak and electromagnetic interactions have little in common. The weak force is approximately a thousand times weaker than the electromagnetic force, and while electromagnetic interactions are long-range, the weak force acts over considerably shorter distances, such as the radius of a neutron. Radioactive decays are also much slower than electromagnetic interactions, and, unlike beta decay, electromagnetic interactions (e. g., an electron scattered by a proton) do not convert elementary particles into different ones. Particles that participate in weak interactions are called *leptons* (from the Greek *leptos*, delicate) and include neutrinos (ν), positrons and electrons ($e^{\perp}$), and muons ($\mu^{\pm}$). They possess very little or no mass.

The weak interaction is related to a fundamental symmetry problem that bears on the concept of time: while electromagnetic interactions are spatially invariant under reflection, the weak force maximally violates the spatial symmetry. Furthermore, the spin of elementary particles plays an important role in weak interactions. Roughly speaking, one may visualize a particle's spin as its intrinsic angular momentum, although spin has no classical analogue according to the correspondence principle. Mathematically, the spin of a particle is represented by a vector parallel to its rotational axis. In compliance with quantum mechanics, a particle's spin can be neither increased nor decreased, and its magnitude for leptons is $h/2\pi$, or simply $^1/_2$. Spin-$^1/_2$ particles can occupy only two spatial orientations: a particle's spin is either aligned with its velocity direction or is opposed to it.

In the same context, one also often refers to a particle's right- and left-handedness, or its *chirality*. For if the right hand is held in such a way that the four fingers indicate the rotational direction of the rotating particle, the right thumb will point in the velocity direction. For left-handed particles, it is the thumb of the left hand that points in the velocity direction.

In 1956, the physicists Tsung-Dao Lee and Chen-Ning Yang proposed certain experiments in which leptons might give preference to a particular handedness. Subsequent experiments did, in fact, show that in "weak decays" only right-handed particles and left-handed antiparticles are emitted. Specifically, neutrinos, which appear to participate only in weak interactions, occur only with left-handed chirality. Antineutrinos are all right-handed. During the beta decay of the neutron (and the muon), only the left-handed portion takes part. In electromagnetic interactions, on the other hand, no particular handedness is preferred and right- and left-handed electrons are represented in equal numbers. Therefore, in weak interactions, spatial reflection symmetry (parity) is violated.

Apart from the symmetry transformations of parity reversal (P) and time reversal (T), the conjugation of charge (C), in which a particle is transformed into its antiparticle, is also recognized. The consecutive execution of all three symmetry transformations P, C, and T defines a symmetry (PCT) that became well known because of the so-called PCT *theorem*, which asserts that the laws of quantum mechanics remain valid upon the combination of parity, charge, and time reversals. This is a trivial result for systems in classical physics, since they are already invariant under each individual symmetry operation. The same also applies to electromagnetic interactions; not, however, to weak interactions.

The reason is that parity reversal P turns a left-handed particle into a right-handed one, which turns out not to occur in nature. To be sure, consecutive P and C transformations turn a left-handed neutrino into its right-handed antineutrino, which does indeed occur in nature. For the weak interaction in beta decay, the product PC is preserved, as is time reversal T, but not P. Hence, one can draw the general conclusion that T-symmetry follows from the PCT theorem together with PC-symmetry. And since virtually all rules governing elementary particles are PC-invariant, they also remain symmetrical in time.

So far only a single case in elementary particle physics is known in which time symmetry is broken. It occurs in the decay of a particle known as the neutral *K-meson*, or *kaon*.[8] Most of the time the K-meson decays into a negative pion, a positron, and a neutrino, a process in which PC-symmetry is preserved; i. e., it also occurs with both parity and charge reversed.

However, in very rare cases (1:1 billion), the K-meson decays into a positive pion, an electron, and an antineutrino, and in that decay, *PC*-symmetry is definitely violated. According to the *PCT* theorem, it would follow that *T*-symmetry is also violated, i. e., the process occurs irreversibly and without time reversal. To be sure, this time irreversibility occurs extremely rarely and is derived only indirectly, since it is based on the universal validity of the *PCT* theorem and on one carefully observed elementary particle decay.

Another symmetry hypothesis was proposed by Steven Weinberg, Abdus Salam, and John C. Ward, who suggested combining the weak and electromagnetic interactions. They made the assumption that in a hypothetical initial high-energy state, weak and electromagnetic interactions are indistinguishable with respect to symmetry transformations. In 1983, at the CERN research center in Geneva, energies high enough to attain these symmetry states were achieved. At certain critical values of energy, symmetry was found to split spontaneously into two component symmetries that corresponded to electromagnetic and weak interactions respectively. This process was explained by the so-called *Higgs mechanism.*

The concept of breaking symmetry is known from many other areas of physics.[9] An example of a system spontaneously losing its symmetry is the transition of a ferromagnet into the magnetized state. As long as the material is unmagnetized, it has no preferential spatial direction, but when magnetized, one spatial axis can be distinguished from all others by the location of the magnetized poles, and the symmetry is broken. The electrons and iron nuclei in an iron rod are then described by equations that are rotation-invariant. The (free) energy of the magnetized rod is invariant with respect to the locations of its north and south poles.

What characterizes the spontaneous breaking of system's symmetry is the magnitude of a control parameter that represents a physical boundary condition of the system, such as its energy. In the context of physical cosmology, symmetry is interpreted as a real condition of the Universe at a particular stage of its development and under the particular temperature and energy conditions that must have prevailed at some time in the past. That means that the Universe itself can be regarded as a high-energy laboratory, whose symmetry conditions can be mimicked to some extent in our terrestrial laboratories.

The *strong force* was originally known as the nuclear force that holds protons and neutrons together in the nucleus. Then, in the 1950s and 1960s, a multitude of new particles were discovered that interacted via the strong force. These particles are now known as *hadrons* (from the Greek *hadros*, strong). Today, the host of hadrons that interact via the strong force is traced back to the symmetry properties of a few *quarks*, particles that constitute the basic building blocks of all hadrons.

Aside from the very special example of the K-meson decay, it appears that elementary particles conform to the time symmetry of a quantum field–theoretical Parmenides world. However, the theories we possess for that world are by no means perfect. Thus, in the relativistic quantum field theories, which unite quantum theory with Einstein's special relativity theory, singularities have cropped up ever since the first attempts at formulating them some 50 years ago, and it is still unclear if these singularities signify fundamental limits of this description of nature. At issue are experimentally measurable quantities, such as the masses of elementary particles and the coupling constants of their interactions, for which quantum field–theoretical calculations yield infinite values. These divergences can be avoided *ad hoc* by the calculation techniques of the so-called *renormalization theories*, but without providing an ultimate explanation.

Carl Friedrich von Weizsäcker has suggested solving these problems of quantum electrodynamics by developing a fundamental logic of time on which a unified physical theory could be based.[10] Starting with simple postulates based on discrete alternatives of time-permitted events (the *ur-alternatives*) that can be selected empirically, he first reconstructs an abstract quantum theory that is valid for all conceivable objects. For the transition from the abstract quantum theory to the concrete quantum theory for real-world objects (e. g., elementary particles) and their concrete dynamics, today's physics generally assumes special, supplemental dynamical laws. Von Weizsäcker, on the other hand, assumes only a single supplemental hypothesis regarding the ur-alternatives (the *ur-hypothesis*), from which one may derive the special and general relativity theories, as well as the quantum theory of elementary particles. The ur-alternatives accordingly describe the binary alternatives on which the state-spaces of the quantum theory can be based. Actually, the ur-alternative represent the information content of a possible binary decision in quantum mechanics as a basic unit. Therefore, an ur-alternative in the sense of von Weizsäcker corresponds to what is now called a *quantum bit*.

Modern elementary particle physics seeks to base the system of elementary particles on specific symmetry groups. Von Weizsäcker begins by demonstrating that his ur-alternatives define a symmetry group that is isomorphic to the transformation group of special relativity theory. The abstract quantum theory derived from the ur-hypothesis then leads to both the existence of three-dimensional real space and the validity of special relativity theory. Until now, however, the derivation of actual particles and fields remains just a program. Its goal is to deduce their existence directly from the ur-theory, while avoiding the divergences of quantum field theories that must presently be evaded by *ad hoc* renormalization techniques. The gauge symmetries of the fundamental physical forces and their particles would have ur-theoretical foundations, according to von Weizsäcker.

In the context of the unification of physics there remains, furthermore, the open problem of the quantum theoretical reconstruction of the gravitational fields of general relativity theory, in which the linearity and delocalization of quantum theory collide with the nonlinearity and localization of Einstein's gravitational equations.

The logic of time is, according to von Weizsäcker, the foundation not only of particles, fields, and the relativistic spacetime continuum, but also, by reason of the concept of probability and statistical mechanics, of the second law of thermodynamics and of the irreversible processes in nature that it deals with. Von Weizsäcker's epistemology is reminiscent of Kant's transcendental philosophy with its insistence on unity and a logic of time as a foundation for understanding the world. It is furthermore evocative of a Platonic natural philosophy: the symmetries of nature are to be derived as approximations of an underlying ur-theory of temporal ur-alternatives.

Time, Black Holes, and the Anthropic Principle

Following the successful unification of electromagnetic and weak interactions (the "electroweak" force), physicists strove for the "grand unification" of the electroweak force with the strong force, and ultimately for the inclusion of gravitation in a culminating step of "superunification" of all four forces of nature. Today, there exist different approaches for achieving this unification, e. g., "supergravitation" or "superstring theory." A direct confirmation of these symmetries is practically out of the question because the demonstration of grand unification alone requires enormous energies. However, it is possible to test the consequences of this theory (e. g., in proton decay). The unification of the natural forces is described mathematically by the inclusion of a growing number of symmetry groups (gauge groups). The great variety of elementary particles and atomic building blocks is the result of broken symmetries in each case.[11]

The beginning of time in the relativistic standard models of cosmology (see Chapter 3) is a singularity that defies explanation. However, in a unified theory of relativistic gravitation and quantum mechanics, the Heisenberg uncertainty relation offers a possible explanation. According to the uncertainty relation, energy and time are interrelated in that the product of the variances of energy and time measurements cannot be less than Planck's constant. That is, the shorter the selected time interval, the greater is the variance in the measured energy. For exceedingly short time intervals, a violation of the energy conservation law becomes possible. Such accidental quantum mechanical fluctuations may have precipitated the initial breaks in symmetry, which led to the rapid expansion of the inflationary Universe.

These considerations also make superfluous the various versions of the *anthropic principle*, according to which the initial conditions of the Universe are deduced quasi-teleologically from the existence of human life.[12] Adherents of this principle argue that the initial conditions were arranged in such a way that after a certain span of time, they would facilitate the evolution of life (and thus a human observer). In contrast to this, the unified theory of quantum gravitation derives the initial conditions of the Universe as a consequence of its own laws (e. g., the uncertainty relation).

The singularity theorems of Roger Penrose and Stephen Hawking also predict that there may be very small regions of space where spacetime is so warped that gravity becomes infinitely great. The existence of such singularities, for example, in the form of black holes, seems increasingly likely today, and several possible candidates, such as the X-ray source Cygnus X-1, have been discovered. In any case, such singularities suffer from the methodological disadvantage that the classical laws of physics are not applicable in regions with infinite curvature, so that it is not possible to predict events in time.

This is why James B. Hartle and Stephen Hawking have suggested a singularity-free model of the Universe, in which quantum theory and general relativity theory are unified and the real time axis is replaced by an imaginary one (in the sense of real and imaginary numbers).[13] In Hawking's unified theory, in contrast to Einstein's classical theory, the three spatial axes together with a complex time axis lead to a closed model of the Universe that lacks boundaries or edges. This spacetime would not only have existed always, but every physical event could be explained in accordance with its laws. In this model, the traditional concepts of everything having somehow "begun" or been "created" are method-ologically simply inappropriate and are revealed as human imaginings originating in our having adapted to the limited spacetime facets of our everyday world of experience.

Hawking's theory is not only mathematically consistent, but is also at least in principle experimentally verifiable; it is therefore a scientific theory and not mere speculation, although it is so far empirically unconfirmed. Among the testable consequences of this singularity-free model is the prediction of black holes in which not all world lines of photons (i. e., light beams) disappear entirely, but are reemitted as measurable amounts of radiation. As in the explanation of the initial singularity of the Universe, the reason lies in the possibility of quantum mechanical fluctuations that are rooted in the uncertainty relation.

But radiating black holes lose energy and mass. In time, they will disintegrate and with them the history of their stars will be lost. In their place, memory gaps will appear in the Universe. With the collapse of its galactic structures, a featureless Universe expanding into a void is heading for a "cosmic Alzheimer's disease." But, perhaps, the laws of quantum mechanics open loopholes (wormholes) of escape from the hopeless cosmic fate. According to general relativity theory, time travel cannot be faster than the speed of light. As light is curved by gravitational fields, time travelers must pass curved paths in spacetime with high speed, limited by the speed of light. Therefore, in order to overcome disruption of spacetime by gravitational fields, spacetime regions would have to be explored using vast curved detours. According to Heisenberg's principle of uncertainty, quantum fluctuations could open short-lived wormholes in spacetime. So, the laws of quantum mechanics make it at least conceivable that wormholes can be employed as fleeting shortcuts between folded regions. However, if our Universe should not be alone, but is instead intertwined with a *multiverse*, along with many other component universes, as was suggested in Andrei Linde's inflationary theory,[14] then wormholes could also be used as escape routes for fleeing a universe that is aging and growing hostile to life.

1   D'Espagnat, *Conceptual Foundations of Quantum Mechanics* (Cambridge, Massachusetts: Perseus Publishing, 2nd ed. 1999); Jammer, *The Philosophy of Quantum Mechanics: The Interpretations of Quantum Mechanics in Historical Perspective* (New York: John Wiley & Sons, 1974).

2   Audretsch and Mainzer, *Wieviele Leben besitzt Schrödingers Katze? Zur Physik und Philosophie der Quantenmechanik* (Heidelberg, Germany: Spektrum-Verlag, 2nd ed. 1996).

3   Everett, "'Relative State' Formulation of Quantum Mechanics," *Reviews of Modern Physics* 29 (1957): 454–62; Wheeler, "Assessment of Everett's 'Relative State' Formulation of Quantum Mechanics," *Reviews of Modern Physics* 29 (1957): 463–65.

4   Penrose and Newton, "Quantum Theory and Reality," *300 Years of Gravity, eds. Hawking and Israel* (Cambridge, Massachusetts: Cambridge University Press, 1987).

5   Misra and Sudarshan, "Zeno's Paradox in Quantum Mechanics," *Journal of Mathematical Physics* 18 (1977): 756.

6   Kwiat, Weinfurter, and Zeilinger, "Interaction-Free Measurement," *Physical Review Letters* 74 (1995), 4763–66.

7   Schwinger, "A Report on Quantum Electrodynamics," *The Physicist's Conception of Nature*, ed. Mehra (Dordrecht, Netherlands/Boston: D. Reidel Publishing Company, 1973), 413–26.

8   Christenson, Cronin, Fitch, and Turlay, "Evidence for the 2π Decay of the K^o_2 Meson," *Physical Review Letters* 134 (1964): 138–40.

9   Mainzer, *Symmetries of Nature: A Handbook for Philosophy of Nature and Science* (New York: Walter De Gruyter, 1996), 441.

10 Von Weizsäcker, *Aufbau der Physik* (Munich, Germany: Hanser, 1985), 390.

11 Greene, *The Elegant Universe: Superstrings, Hidden Dimensions, and the Quest for the Ultimate Theory* (New York: W. W. Norton & Co., 1999); Mainzer, *Symmetries of Nature: A Handbook for Philosophy of Nature and Science* (New York: Walter de Gruyter, 1996), 464.

12 Barrow and Tipler, The Anthropic Cosmological Principle (New York: Oxford University Press, 1988).

13 Hartle and Hawking, "Wave Function of the Universe," *Physical Review* D31 (1938); Hawking, *A Brief History of Time: From the Big Bang to the Black Holes* (New York: Bantam Doubleday Dell Publishers, 10th anniversary ed. 1998).

14 Linde, *Inflation and Quantum Cosmology* (New York: Academic Press, 1990); Linde, A., Linde, D., and Mezhlumian, "From the Big Bang Theory to the Theory of a Stationary Universe," *Physical Review* D 49 (1994): 1783–1826.

chapter 5

Time and
Thermodynamics

Our everyday lives seem to be controlled by a fundamental asymmetry of nature: The past and the future are not interchangeable, youth will not return, and the dead will not come back to life. Life appears to run preferentially in one direction. Humans have been preoccupied with the meaning of this "time arrow," which appears to determine their fates, ever since the beginnings of philosophy and religion. Since the end of the nineteenth century, this problem has also been tackled with mathematical and physical precision within the framework of thermodynamics. Entropy may be interpreted statistically as a measure of disorder in systems such as a collection of gas molecules in an insulated container. The spontaneous increase of disorder, as well as the decay of ordered systems, is predicted by the second law of thermodynamics. But how are we to understand cosmic evolution, which appears to develop ever more complex ordered systems from simpler ones?

Time in Equilibrium Thermodynamics

Our physical observations appear to confirm nature's time arrow. A mechanical watch shatters into its components when it is dropped on the floor. The temporal reversal of this mechanical event, namely the spontaneous reassembly of screws, springs, and gears into a running watch, has not been observed. The observation of electrodynamic phenomena also appears to exclude time reversal. Thus, radio stations and stars emit spherical electromagnetic waves, but the inverse process, the concentric reception of radiation from all directions, has so far not been observed.

The properties of heat also reveal the existence of a specific time direction. According to the second law of thermodynamics, heat in an isolated system (i. e., a system in which no material or energy exchange with its surroundings occurs) always distributes itself in such way that a certain system parameter (*entropy*) is never diminished but always increases or remains constant. Entropy is interpreted as the degree of disorder, so that a well-ordered state has lower entropy than a chaotic one. A pot of hot coffee cools spontaneously to room temperature, while room temperature in turn increases slightly. The inverse process of the coffee spontaneously warming up relative to room temperature has not been observed. Heat continues to flow in a system until it is uniformly distributed everywhere and the system no longer contains any temperature gradients. Only in that final state of thermal equilibrium is the summit of a system's temporal development attained.

It is remarkable that the arrow of time, which is obviously so familiar from our everyday perceptions, is very difficult to understand physically. Historically, it has been investigated relatively late, as the time reversibility of the equations of motion has seemed more surprising. Thus the first of Newton's laws (the law of inertia) is reversible in the sense that in the absence of forces, every solution of the equation for time t has a corresponding solution for time $-t$. The time reversal expresses itself as a sign change in the direction of motion. The reversibility of motion is therefore manifested by a reversal of the order in which positions are occupied.

Huygens's law of pendulum motion offers a simple example. According to the law of conservation of energy, a pendulum would continue swing to and fro indefinitely as it converts potential energy into kinetic energy and vice versa. Its motion is therefore always reversible, and hence we can calculate the position of the pendulum in the future as well as in the past. In that sense, no specific direction of time is selected. But in reality, the pendulum continuously loses kinetic energy, which is converted irreversibly into heat by friction, so that the pendulum eventually comes to rest.

Secondary conditions like friction can be traced to irreversible thermal processes, as indicated above. The process of our coffee pot cooling to room temperature is described by a heat flow equation, in which time reversal $(t \rightarrow -t)$

causes the sign of the heat conduction coefficient to be changed from positive to negative. The time symmetry of the dynamics is thus broken in this example.

In the 1860s, Rudolf Clausius introduced a "conversion value" of heat, which was intended to characterize the irreversible processes in an isolated system by its spontaneous increase.[1] In analogy to the Greek word *energeia* for "energy," Clausius coined the word "entropy" from the Greek word *tropos*, meaning "change." The change in entropy S of an open physical system and its environment at time t is given by the sum $dS = d_eS + d_iS$, where d_eS represents the change in entropy due to an exchange with the surroundings, and d_iS is the change in entropy within the system itself. The second law of thermodynamics then requires that $d_iS \geq 0$ for isolated systems, for which $d_eS = 0$. It follows that for isolated systems entropy either increases or else it stays constant once thermodynamic equilibrium is attained, i. e., when $d_iS = 0$.

In the beginning thermodynamics was primarily a phenomenological theory for describing the distribution of heat in macroscopic bodies. Ludwig Boltzmann suggested a statistical mechanical explanation by relating a body's macroscopic conditions, such as heat, to the kinetic properties of molecules.[2] Boltzmann considered in detail the momentum carried by individual gas molecules and their effect on the statistical distribution functions for the position and velocity of the molecules. In 1872, he introduced an entity H to characterize the statistical distribution of the molecules in velocity space, defining it as the average of the logarithms of the distribution function obtained by integrating over velocity. He initially called this quantity E, which became H for Greek-capital "eta," and eventually, today's term: H. Boltzmann was then able to prove that the quantity H remains constant if the velocities of the molecules follow the Maxwell distribution. Otherwise, H always decreases, as is the case in diffusion, friction, or heat conduction.

The quantity $-H$ is in that sense analogous to the entropy S. The idea behind Boltzmann's statistical-mechanical supposition is that macroscopic events can be explained in terms of microscopic processes involving very many particles with very many degrees of freedom.

As an example of a macroscopic process, consider a container filled nonuniformly with a gas, which very quickly reaches a state of constant density. Such a conversion of a macroscopically nonuniform distribution into a macroscopically uniform distribution takes place with high probability, while the reverse process is extremely unlikely, as is demonstrated by the following simple example: Consider the possible ways in which $N = N_1 + N_2$ particles can be distributed between two identical boxes, with N_1 in the left box and N_2 in the right. If the total number of particles is $N = 10$ and the particles are distributed randomly between the two boxes, then if the unlikely distribution (10.0) occurs on the average once, the distribution (9.1) will occur 10 times, the distribution (8.2) will occur 45 times, and so on. The most likely equilibrium distribution (5.5) will occur 252 times. These number are readily derived for $N = 10$ from the following formula for W, the number of ways in which the particular distribution (N_1, N_2) is obtained in distributing $N =$ particles:

$$W = \frac{N!}{N_1! N_2!}$$

where $N!$ ("N factorial") is defined by $N! = 1 \times 2 \times 3 \times \ldots \times N$, with $0! = 1$.

Statistical mechanics explains macroscopic conditions such as spatially varying density, pressure, or temperature by microscopic events, and, as a result, we can say that an observable macroscopic state is realized through a large number W of microscopic states. In determining the number W, one considers a large number of independent, equivalent mechanisms and particles such as atoms, molecules, liquids, crystals, etc. These pass through their microstates on the basis of equations of motion with their different initial phases. When a particular macrostate is realized as a result of W such microstates, its entropy is given by $S = k \ln W$, where k is Boltzmann's constant. According to Boltzmann, the entropy of a system is a measure of the probability that the molecules arrange themselves in such a way that the system finds itself in the observed macrostate.

In 1875, Lord Kelvin and Joseph Loschmidt raised a fundamental "reversibility objection" to this idea, which has a direct bearing on the problem of time symmetry. How, they asked, can Boltzmann's H theorem, which includes irreversible processes, be derived from the principles of mechanics (i. e., from the equations of motion of microscopic particles), which are invariant with respect to time

reversal? Boltzmann answered this by stating that the second law of thermodynamics is based not only on mechanics, but also on the subsidiary assumption of an extremely unlikely initial condition. If one begins with an unlikely initial distribution, it will subsequently most likely be converted into more uniform distributions. However, many more uniform distributions are likely to be converted into similarly uniform distributions, so that in most cases time reversal of the microscopic motions has no effect on a uniform distribution. The second law is therefore valid with very high probability but not with certainty. Irreversible processes are accordingly merely frequent or probable ones, while their reversals are rare and unlikely processes.

Local deviations or fluctuations are permitted by the second law, but Boltzmann did not live to see their experimental confirmation. In 1905, Einstein showed that fluctuations indeed occur in nature and that they originate in local violations of the second law's overall probability trend. He demonstrated this by the example of Brownian motion, a phenomenon that had long been known to botanists.[3] In this experiment, microscopic particles suspended in a liquid are observed to move with an irregular quivering motion as a result of being pushed this way and that by the random impulses imparted by liquid molecules, although the most likely outcome is for the numerous impulses to compensate each other completely.

The "revisiting objection" raised by Jules Henri Poincaré and Ernst Zermelo (1896) maintained that every state of a mechanical system with a finite number of degrees of free-

dom must return (approximately) to a particular state after a certain time.[4] Accordingly, all states of the system are revisited, at least approximately, and consequently a time arrow based on increasing entropy cannot exist. To this, Boltzmann responded that—as the number of degrees of freedom increases—the times of return become exceedingly long.

Boltzmann's insights led to two points of view regarding the conclusion that the laws of mechanics are time-reversible, but real events are irreversible:

1 the world originated from a highly improbable initial state; or
2 as long as the world is sufficiently large, strong deviations from uniformity must exist somewhere.

In the generation and dissolution of such strong deviations, the process proceeds in a unique direction, and we perceive this as the time arrow.

Boltzmann's fluctuation hypothesis starts with the supposition that the entire Universe is in thermal equilibrium, and hence in a state of maximum disorder. However, it is assumed that local fluctuations in entropy exist in such a Universe; (i. e., there are regions of spacetime that contain ordered structures). According to Boltzmann, the two time directions of the Universe are considered to be completely symmetrical. Consequently, entropy increases in both time directions in a similar way before leveling off at the maximum entropy values. Since entropy increases in both time directions from its minimum, it is as likely that we are living in a phase of increasing entropy as of decreasing entropy.

The validity of the H theorem would then be mere coincidence. Boltzmann defended his ideas by stating that no objectively selected time direction exists; instead, it is merely so perceived in a particular world. People would then measure time in either of the two branches "in the direction of increasing entropy." A clarification (although not a solution to all problems) was provided by the work of Tatyana and Paul Ehrenfest, who used simple and completely transparent models to simulate the irreversibility of complex, real processes, which often are not comprehensible in detail.[5]

The (small) deviations of thermodynamic variables from their equilibrium values have quadratic dependences, as was demonstrated by Josef Meixner.[6] This is characteristic of entropy changes and expresses the system's symmetry with respect to the equilibrium point. Since all deviations of entropy from its equilibrium value have negative signs, entropy can only increase when equilibrium is reestablished, independent of the direction of time. This explains why near equilibrium, entropy changes can no longer be used to distinguish between past and future.

Because humans have an awareness of time, Boltzmann's successors concluded that the region of the Universe that we inhabit is still far removed from an equilibrium state. Nonequilibrium thermodynamics could provide the appropriate physical basis for a theory of life.[7] Although this branch of thermodynamics was nineteenth century, Boltzmann nonetheless had brilliant cosmological insights regarding the connection between thermodynamics and

biology that formulated for the first time the connection between the asymmetry of time and the evolution of life.

Time in Nonequilibrium Thermodynamics

A system is in thermodynamic equilibrium with its environment when its macroscopic, collective properties (such as pressure and temperature) are exactly the same as its environment's. As an example, consider a layer of liquid between two horizontal parallel plates. Left on its own, the liquid will approach thermodynamic equilibrium, a homogeneous state in which the molecules or particles of the liquid are statistically indistinguishable. The system is then in a state of complete symmetry in which no macroscopic changes take place, and there are no temperature differences with respect to the outside world. If T_1 and T_2 are the temperatures of the upper and lower plate, respectively, then $\Delta T = T_2 - T_1 = 0$.

If the system is perturbed by the warming of the lower plate, so that $\Delta T > 0$, it will return to its equilibrium state of its own accord as long as the temperature differences are small. But if ΔT increases, and the system is driven from its equilibrium state, new macroscopic features suddenly appear. In fact, the liquid arranges itself into small regular cells inside of which the layers rotate. This so-called *Bénard convection* is caused by the rising and falling currents initiated by density variations in the liquid next to the differentially warmed plates. It reflects a genuine break in spatial

symmetry: the liquid in the convection cells rotates alternately to the left and the right and thereby selects a preferred direction in each cell.

Such nonequilibrium situations are plentiful in nature. For example, the biosphere is generated by the radiative equilibration between the Sun and Earth. Complex systems that are far from thermal equilibrium generate new structures and properties spontaneously, which can emerge as new states. The temporal development of complex systems beyond their threshold values can be visualized by the so-called bifurcation diagrams, which depict the dependence of a system's state parameter s on a control parameter λ (see Figure 5.1). In the example of Bénard convection, thermodynamic equilibrium is maintained for small values of λ (in this case, for small values of ΔT); the liquid is in a state of asymptotic stability and quenches internal fluctuations on its own. When λ exceeds a critical threshold value λ_1, this development branch of the liquid becomes unstable. The system can no longer quench its own oscillations, and in the Bénard experiment the liquid begins to rotate either to the left or to the right. Bifurcations in which generate new development branches are geometric representations of symmetry breaking that is characteristic of complex systems.[8]

The dynamical development of the states of a complex system is described by a nonlinear differential equation that is a function of the control parameters. Nonlinearity refers to the mathematical form of functions. For linear functions, the function values change proportionally to the function arguments. Therefore, in linear dynamical systems, the

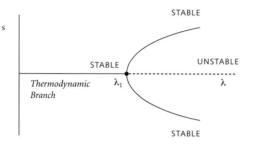

FIGURE 5.1 Bifurcation diagram of the thermodynamic branch.

change of effects is proportional to the change of their causes. The term *linearity* refers to the geometrical representation of linear functions: In geometry, linear functions represent straight lines, nonlinear functions do not. Examples of nonlinear functions are polynomial or trigonometric functions. In nonlinear dynamical systems, the change of effects is no longer proportional to the change of causes. A nonlinear temporal development can lead to new ordered structures, as in the example of Bénard convection, but it can also lead to chaos. This can be illustrated by the logistic function that has also been used to model the growth of animal populations, or the spreading of epidemic diseases. In order to calculate growth rates at successive time intervals (e. g., days), the growth rate at time $t + 1$ is calculated from the growth rate at time t, according to the formula of the logistic function.

Such recursion formulas generally tell us no more than how new values may repeatedly be calculated with the previously calculated values according to the given functional prescription. Thus, they represent feedback in the dynamic system. In the case of the logistic map, the new value x_{t+1} is calculated with the previously calculated x_t according to the quadratic, nonlinear equation $x_{t+1} = \lambda\, x_t(1 - x_t)$, generating a sequence $x_1, x_2, x_3, \ldots$ of values for $t = 1, 2, 3, \ldots$ The system's temporal development as a function of the control parameter λ may again be represented by a bifurcation diagram with the typical symmetry of breaking equilibrium states (Figure 5.2a). When the system's temporal development is represented in an appropriate way, its state curve is found to split into 2 branches (bifurcation) at a λ-value of λ_1. It splits into 4 branches at $\lambda = \lambda_2$, into 8 branches at $\lambda = \lambda_3$, and generally into 2^n branches at $\lambda = \lambda_n$. The latter generates a *periodically doubled bifurcation cascade.* Up to a critical value λ_c, the limiting value of the series $\lambda_1, \lambda_2, \lambda_3, \ldots$, it is possible to predict the development of this tree and its corresponding dynamic system. For values of λ that exceed a certain λ_c, however, the bifurcation tree turns into a dense gray pattern of points (Figure 5.2b). Although each point is still uniquely determined by the logistic equation, the behavior of the system is no longer predictable: chaos reigns.[9]

Nonlinear systems may be dissipative or conservative. Dissipative dynamic systems are those that consume matter (e. g., energy or human labor) and then release it to their environment as heat. Living organisms, in particular, are open dissipative systems that exchange matter and energy with their environment, generating new forms and structures. The complex interactions of the system's components produce synergistic effects that may lead toward new forms and structures by self-organizing processes (phase transitions), or they may lead to chaos.

Dynamic systems are called conservative if they are closed with respect to their environment. They include almost all the (idealized) frictionless systems of classical mechanics, in which energy exchanges take place only within a closed system. The interchange between kinetic and potential energy in an ideal pendulum and ideal frictionless planetary motion are two typical examples. Leibniz, Newton,

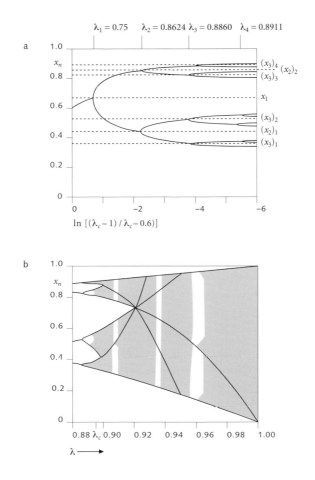

Figure 5.2 Sequence of period-doubling bifurcations (a) and chaotic regime of the logistic map beyond λ_c (b).

(From Mainzer, Thinking in Complexity: The Complex Dynamics of Matter, Mind, and Mankind *[New York: Springer-Verlag, 3rd ed. 1997], 62.)*

and Einstein were in agreement that in such systems, causes and effects are not only uniquely connected with each other, but may in principle be calculated to arbitrary precision. This deterministic worldview was based on faith that—with the aid of the equations of motion and sufficient information regarding the system—the future could be predicted as far ahead as desired, and to any precision. The belief in an omnipotent computability of nature became famous as *Laplacian spirit*, after the French astronomer and mathematician Pierre Simon Laplace.

Today, we know that this assumption is not universally valid, even in classical mechanics. Indeed, even the simple case of more than two bodies interacting with each other can lead to orbits that have a strong dependence on the initial conditions. A miniscule variation (e. g., a tiny perturbation of a planet's orbit by a comet), which is initially deemed to be negligible, can intensify chaotically and bring motions into play that are no longer calculable. At the end of the nineteenth century, Henri Poincaré had already encountered the chaotic behavior of the deterministic laws of astronomy and had shown mathematically that the three-body problem is not integrable and can lead to utterly chaotic orbits.

Doubts now began to appear about whether deterministically predicted temporal developments are in principle calculable. But it took the power of modern supercomputers to elucidate these limitations persuasively. The deterministic world picture, with its belief that everything is completely calculable and achievable, proved to be an illusion. Instead, reality seems to be shaped by complex structures and by self-organizing processes that may either stabilize or dissipate suddenly.

Such self-stabilizing structures are the so-called *attractors*, to which the states of the complex system are drawn under particular secondary conditions. Geometrically, the state of a complex system can be represented by a point of the system's phase space. The time-dependent development of the system is an orbit (*trajectory*) of sequential points representing sequential states in the phase space. Consider, for example, the dynamics of a population according to a logistic function: in the case of weak growth with low values of the corresponding control parameter, all trajectories are drawn to a certain point of equilibrium (*point attractor*) in the phase space. If the growth increases, the system starts to oscillate periodically between two points of stability, and all trajectories are drawn to a limit cycle in the phase space. In the case of very strong growth beyond a critical value of the control parameter, the system becomes chaotic, and all trajectories are drawn to a bounded region of irregular, nonperiodic behavior in the phase space (*chaos attractor*).

More than 60 years after Poincaré, the work of Andrey Nikolaevich Kolmogorov (1954), Vladimir Igorevich Arnold (1963), and Jürgen Moser (1967) produced a general proof of the so-called *KAM theorem*, which states that in the phase space of classical mechanics, motion is neither completely regular nor completely irregular, but that orbital trajectories depend sensitively on the selected initial conditions.[10] Stable regular classical motion turns out to be the exception.

An example of dissipative chaotic systems is provided by the nonlinear differential equations used by meteorologists. According their calculations, a global chaotic change in the overall weather pattern could be precipitated by the tiniest local change a small unnoticed turbulence in the weather map, a fluttering leaf, the beating of a butterfly's wing. We all know about the reliability of weather reports, and following the lead of American meteorologist Edward Lorenz, the term "butterfly effect" is indeed used in mathematical chaos theory. The uniqueness (in principle) of the solution to mathematical equations cannot guarantee the solutions' capability to predict future events arbitrarily far ahead and with arbitrary precision. Ironically, it is the great computational power of modern computers that provided an unambiguous demonstration of these limitations.

Time, Irreversibility, and Self-Organization

The nonlinear temporal development of dynamic systems can lead not only to chaos, but also to new ordered systems by a process of self-organization. Accordingly, many phenomena that have historically been considered to be irreducible properties of living organisms (self-organization, metabolism, spontaneity, emergence, Gestalt, etc.) are now verifiable and explainable within the framework of physics and chemistry. New ordered structures are generated far from thermal equilibrium as a result of variations in certain external control parameters (temperature, imported energy) until the former state becomes unstable and switches to a new state. These phase transitions can be understood as breaks in the symmetry of equilibrium states. At critical values of a control parameter, macroscopic ordered structures appear spontaneously. These ordered structures are the outcome of collective (synergistic) cooperation among the system's microscopic components. Consequently, the creation of order (emergence) is by no means improbable or accidental, but takes place under specific conditions and follows specific rules.

Consider, for example, the flow pattern of a river passing an obstacle (e. g., a bridge pier) as a function of the flow velocity. At first, for low velocities, the river retains a homogeneous flow pattern past the obstacle; i. e., it strives toward a homogeneous equilibrium state as its attractor. But as the flow velocity increases, oscillations in the form of eddies are generated. At first periodic bifurcations are generated, then at higher velocities quasi-periodic oscillations are formed, and finally the eddies are transformed into a chaotic and fractal pattern. The interactions of the liquid's molecules on the microscopic level, which depend on the flow velocity as the control parameter, have now generated a new macroscopic flow pattern.

A famous example of such a phase transition is the spontaneous generation of coherent laser light by the synchronization of originally unrelated photons. This happens when the externally supplied energy has reached a certain high critical value.[11] In meteorology, the spontaneous appearance of cloud formations may be described in terms of phase transitions that occur when temperature and other environmental parameters reach certain critical values. In chemistry, liquids are known to form patterns called *dissipative structures* that are generated by the addition of energy-rich substances to particular mixtures. These structures can be maintained as periodic pulsations (chemical clocks). In all of these cases are the cooperative effects among countless molecules responsible for the phase transitions that lead to new ordered states.

In the so-called *Zhabotinsky reaction,* circular waves are generated spontaneously on the surface of a chemical mixture. These break up into spiral waves that eventually displace all circular waves, although such circular waves ought to penetrate unperturbed according to the superposition principle. This is a direct illustration of the nonlinearity of complex systems and of the limitations of the superposition principle.

In cosmology, chaos theory and the self-organization theory are able to elucidate how small fluctuations in the distribution of matter are amplified in the course of time. These small fluctuations cause initially uniform clouds of cosmic gas to break up into galactic islands, and eventually lead to the creation of billions of individual stars. The beginning of the cosmos was characterized by a high degree of symmetry and order and hence low entropy. If the second law of thermodynamics is generalized to the Universe, it predicts the increase of entropy, which eventually ends in a thermal death of maximum disorder. This scenario is consistent with the relativistic standard models, which predict an unlimited expansion of the Universe (see Chapter 3). According to the closed model, gravity will, at some point, overcome the expansion force and lead to the reversal of the Universe's expansion, back to the ultimate singularity of cosmic collapse. However, not all processes run backward once that threshold is reached. Entropy continues to increase even after the cosmic collapse has begun, so that the second law remains valid. Evidently, time is not identical with the development of entropy, so that the initial and final singularities of this model are by no means symmetrical. But then the question arises of how to explain the asymmetry of time and the irreversibility of cosmic evolution with its increasing number of irregularities (e. g., black holes), from its initial singularity with low entropy to a final high-entropy singularity.

Irreversible, nonlinear processes of development ought to be derivable in the framework of a unified theory of quantum gravitation. Penrose sees a connection between the transition from the reversible, nonlocal states of the quantum world to the macro-world's localized states and a significant curvature of cosmic spacetime due to gravitational effects. But gravitation only becomes effective for macroscopic systems that exceed a certain size.

In an effort to comprehend the temporally asymmetrical development of the states of a dynamic system in thermodynamic terms, Ilya Prigogine draws a distinction between the reversible "exterior" time of a system and its irreversible "interior" time, or its "age." While the "exterior time" is the usual real time parameter *t* that is registered by a clock, the "interior" time is defined as an operator that takes into account the irreversible changes in the system's states.[12] As a real parameter, the exterior time appears merely as an index in a unique set of trajectories (in classical physics) or in a wave equation (in quantum mechanics). As an operator, the interior time permits statements about the temporal development of a complex collection of trajectories or distribution functions that serve mathematically as the eigenfunctions of the time operator. The connection with external time rests on the eigenvalues of the time operators being real lengths of time as registered by a normal clock. The distributions are graphical representations of the different interior "ages" of a complex system. For example, the different organs of a complex system like the human organism wear out at different rates. The time operator assigns a "mean age" to each state of the system, which increases at the same rate as the exterior clock time.

From the viewpoint of the history of philosophy this is reminiscent of Aristotle, who distinguished between time as "movement" (*kinesis*) and time as "coming into being, growth, and decay" (*metabole*). Prigogine connects this distinction to the concepts of reversible time in mechanics and of irreversible time in thermodynamics. Irreversible processes are explained according to the second law of thermodynamics as internal breaks in symmetry (based on the time operator) that violate time reversal symmetry. Prigogine's time operator has the remarkable property that the past and future are separated by an interval that is quantifiable in terms of a characteristic time. Traditionally, the present is represented as a point on the time axis in which past and future can come infinitely close. Prigogine therefore speaks of the "duration" of the present, which he compares to Henri Bergson's concept of duration (see Chapter 7). The time operator is, however, a mathematically defined system parameter that must not be confused with subjectively experienced time.

1   Schneider, "Rudolph Clausius' Beitrag zur Einführung wahrscheinlichkeitstheo-
    retischer Methoden in der Physik der Gase nach 1856," *Archive for History of
    the Exact Sciences* 14 (1974/75): 237–61.

2   Boltzmann, *Über die mechanische Bedeutung des Zweiten Hauptsatzes der
    Wärmetheorie* (New York: reprint ed. 1968), 9–33.

3   Einstein, "Über die von der molecularkinetischen Theorie der Wärme geforderte
    Bewegung von in ruhenden Flüssigkeiten suspendierten Teilchen," *Annalen der
    Physik* 17 (1905): 549–60.

4   Poincaré, "Sur les tentatives d'explication méchanique des principes de la ther-
    modynamique," *Comptes rendus de l'Academique des sciences* 108 (1889):
    550–53; Zermelo, "Über einen Satz der Dynamik und die mechanische
    Wärmetheorie," *Annalen der Physik* 57 (1896): 485.

5   Ehrenfest, "Zur Entropiezunahme in der statistischen Mechanik von Gibbs,"
    Wiener Berichte 115 (1906): 89.

6   Meixner, "Die thermodynamische Theorie der Relaxationserscheinungen und
    ihr Zusammenhang mit der Nahwirkungstheorie," *Kolloid Zeitschrift* 134
    (1953): 3.

7   Prigogine, *Non-Equilibrium Statistical Mechanics* (New York: J. Wiley & Sons,
    1962); Haken, *Synergetics: Nonequilibrium Transitions and Self-Organisation
    in Physics, Chemistry, and Biology* (New York: Springer-Verlag, 3rd ed. 1983).

8   Nicolis and Prigogine, *Die Erforschung des Komplexen* (Munich, Germany:
    1987); Mainzer, *Thinking in Complexity: The Complex Dynamics of Matter,
    Mind, and Mankind* (New York: Springer-Verlag, 3rd ed. 1997).

9   Mainzer and Schirmacher, *Quanten, Chaos und Dämonen:
    Erkenntnistheoretische Aspekte der modernen Physik* (Mannheim, Germany: BI
    Wissenschaftsverlag, 1994), 38.

10 Arnold, "Small Denominators II. Proof of a Theorem of A. N. Kolmogorov on
    the Preservation of the Hamiltonian," *Russian Mathematical Surveys* 18
    (1936): 5; Kolmogorov, "On the Conservation of Conditionally Periodic
    Motions for a Small Change in Hamilton's Function," *Dokl. Akad. Nauk. USSR*
    98 (1954): 525; Moser, "Convergent Series Expansions of Quasi-Periodic
    Motions," *Mat. Ann.* 169 (1967): 163.

11 Haken, "Laser Theory," E*ncyclopedia of Physics* XXV/2c
    (Berlin/Heidelberg/New York: 1970).

12 Prigogine, *From Being to Becoming: Time and Complexity in Physical Sciences*
    (New York: W. H. Freeman, 1981), Chapter X.

chapter 6

Time and Life

The thermodynamic concept of time is directly applicable to the discussion of life processes. Charles Darwin and Herbert Spencer's theory of evolution for the first time related growth and life to developing complexity. The evolution of life is revealed as an irreversible temporal development of complex systems. In the framework of nonequilibrium thermodynamics, evolution can be understood in terms of symmetry breaking. This lies at the root of life's time arrow. But many different biological temporal rhythms must be distinguished. In the course of evolution, these rhythms were superimposed on each other into complex time hierarchies, encompassing everything from elaborate economic systems to single living organisms.

Time in Darwin's Theory of Evolution

Historically, time and life have always been perceived as being closely connected. As was mentioned before, the Aristotelian tradition distinguished between time as movement and time as becoming and decay. In biology, the word "evolution" was used primarily in the so-called *pre-formation theory*, probably beginning with the work of the Swiss scientist Albrecht von Haller in 1744 and continuing on into the nineteenth century. That theory was based on the supposition that the structures of the completed organism are already present in the egg and sperm and that the components merely unfold in all subsequent developmental phases. This kind of theory of "evolution" or "unfolding" required no new act of creation subsequent to the Creation and was readily compatible with the literal biblical accounts. It competed with the so-called *epigenetic theory*, in which the emergence of complex structures is not predestined, but is made possible by an act of "*creatio ex nihilo.*"

This controversy regarding embryonic development was still very much alive at the time Darwin published his momentous work *On the Origin of Species* (1859). To prevent any erroneous analogies to embryology, Darwin attempted to explain the formation of new species from old ones while avoiding the use of the term "evolution" in his book. Writing about fossils in 1832, Sir Charles Lyell, a British geologist, was the first to use "evolution" in the modern sense of the word. The generalization of the concept of evolution to all development processes of living organisms is largely due to the work of Herbert Spencer, for whom evolution meant progress toward ever-greater complexity.[1]

Darwin's conclusions regarding the development of biological species by natural selective breeding seemingly made superfluous the assumption that life and nature are guided by teleological forces. Around 1900, Ludwig Boltzmann sketched out a completely reductionistic model of life which was based on nineteenth-century thermodynamics and his knowledge of physics and chemistry. In this model, he anticipated many of today's scientific views about life. For example, Boltzmann raised the question, How is it possible that nature, ostensibly programmed to head for disorder, death, and decay in accordance with the second law of thermodynamics, actually proceeds toward ever more complex ordered structures and living systems?

From the standpoint of equilibrium thermodynamics, the development of life is an instance of swimming against the current of entropy, which seeks to eradicate all order unless the expenditure of energy counters its effects. The alternative—the spontaneous creation of order in the absence of an external energy supply—would contradict the second law and would require "demonic" forces.

The idea of such a demon—one that is capable of preventing the irreversible increase of the entropy of a closed and isolated system in conformity with the second law—would constitute a *perpetuum mobile* of the second kind, a concept first proposed by James Clerk Maxwell. In 1879, Sir William Thomson (later Lord Kelvin) first mentioned the "sorting demon of Maxwell," which was able to separate faster gas molecules from slower ones in two connected glass containers. By doing so, one could cause the first container to be warmed spontaneously, while the other was cooled.[2] Although Maxwell's demon is merely a participant in a *Gedankenexperiment*, it was interpreted as a paradox of thermodynamics for many years. But if the demon is not considered to be all too wondrous a creature, and in fact needs to metabolize in order to do its work, the concomitant entropy production could make up for any entropy deficit related to its sorting activities. In that case, the generation of order by Maxwell's demon is accomplished at the cost of the expended energy and cannot be considered a violation of the second law.

But strictly speaking, the second law of thermodynamics is not applicable to life processes. For living systems are examples of open systems that stubbornly avoid thermal equilibrium and decay by continually exchanging matter and energy with their environment through metabolic processes. While thermodynamics after Boltzmann was focused on equilibrium situations, life manifestly takes place far from thermal equilibrium. A mathematical and physical nonequilibrium theory was not formulated until much later (see the references for Prigogine and Haken in Chapter 5), and it eliminated the need for Maxwell's demon as the creator of order.

Nonlinear feedback mechanisms facilitate the flow of energy and matter that is needed to create and maintain functional and structural order. This means that new structures are the result of dissipative and conservative self-organization. Since these processes are irreversible, they also represent the interior time of evolution, and in that sense, life is the result of temporal symmetry breaking.[3]

Time in Molecular Evolution

There is a striking analogy between the basic concepts of biological evolution and nonequilibrium thermodynamics. The emergence of new biological forms is analogous to the establishment of thermal equilibrium, with mutations corresponding to fluctuations. A complex dynamic system's quest for stability is achieved in biological selection, and the branch points of bifurcation diagrams are similar to the family trees describing biological evolution. In the complex physical and chemical systems studied so far, the self-organization resulting from the phase transitions at bifurcation points is the result of selection. In a laser pulse, for example, the wave trains emitted by the individual excited atoms have ceased to compete and have spontaneously adopted a common phase and direction. In this sense, selection again signifies symmetry breaking.

To illuminate the transition from inanimate matter to animate nature, mathematical equations for modeling evolution have been suggested. These equations describe the development of biomolecules as a process of self-organization. They assume the existence of self-amplification by autocatalytic processes that Manfred Eigen and Peter Schuster have described in their models of hypercycles.[4] In this model life is created as the result of successive self-optimizations of a molecular system, reached by a sequence of intermediate selection steps. Life is not, as Jacques Monod asserted, the result of a solitary accidental event, a unique singularity, in which the phase state of inanimate matter is destabilized as a result of a random fluctuation and switches spontaneously to a new equilibrium state, which we call life. According to Eigen, the creation of life (in terms of mathematical catastrophe theory) is not the result of a single spontaneous break in symmetry, but rather, is the result of a series of local symmetry breaks, in which unstable selection equilibria are replaced by new and more "valuable" equilibria.

The boundary between inanimate and animate nature is accordingly a fluid one. In fact, selection and self-organization processes take place even in inanimate matter and can be described physically in terms of extrema (variational) principles. In the case of particular macromolecules capable of biochemical autocatalysis, these principles inevitably lead to developments that lay the foundations for life. From a theory of science point of view, known principles from physics and chemistry are all that are needed. To be sure, the use of variational principles makes it possible to describe the process in teleological language, which speaks of developments that are steered by "goals" and "purposes," but this in no way implies the existence of vitalistic forces in inanimate nature.

In the successive self-optimizations of molecular systems, it is possible to ascertain the gradient of the evolutionary direction as the mathematical expression of the time arrow, but not to predict all future evolutionary events. On the other hand, a precise *retrospective* analysis of temporal bifurcation diagrams is possible. For instance, by determining the molecular sequences that occur in genes of particular species, one may draw conclusions about the relationships of different organisms and their historical evolution. Apart from the traditional methods of paleontology and comparative morphology used since Darwin's time, considerably more precise methods are now available for testing theories of evolution on the molecular level. It is thus possible to use the antecedents of related genes to determine the respective archetypal genes and to construct an ancestral tree of molecular evolution. In a certain sense, this amounts to a genetic clock, in which mutations represent the measure of age. It is therefore in principle possible to reconstruct the evolution of organisms, from bacteria to primates, in terms of temporal bifurcation diagrams.

It is a remarkable symmetry property of these evolutionary processes that the invariability of a species can be preserved over many generations by the invariance of the DNA structure. Only when mutations appear and the prevailing selection equilibrium becomes unstable does symmetry breaking occur, which is revealed macroscopically in biological structures and forms.

As convincing as Eigen's model of evolutionary self-replication and selection is for the description of life processes that are observable today, it nevertheless fails to provide a complete explanation for the origin of life. The reason is that to explain replication and self-reproduction, Eigen presupposes the existence of an information processing mechanism that is both exceedingly simple and functions with high efficiency and a low error rate. It remains to elucidate how such nearly perfect molecular machinery for self-replication could have emerged in the early phases of life. In other models, molecular self-replication and metabolism come into existence not simultaneously, but consecutively. But whatever the case may be, biological evolution can be properly addressed by the theory of open complex systems far from thermal equilibrium, whose development is described by nonlinear equations just like those for laser light or chemical reaction rates.[5]

The path of evolution leads by way of cell differentiation[6] to organisms, and mathematical models that simulate the development of these new structures are already available.[7] In a cluster of cells, the stimulatory and inhibitory agents for generating its eventual form are initially uniformly distributed. Interactions between the cells lead to chemical reactions and diffusion processes that cause the production rate of, say, the stimulant, to reach a critical value. As a result, a chemical pattern is generated with density gradients that stimulate the cellular genes to develop the cells' diverse functions.

Analogous models based on the theory of complex dynamic systems are also applicable to systems of organisms, such as the populations of plants, bacteria, or animals. In fact, nonlinear equations were used to model biological populations as early as the nineteenth century. Thus, the nonlinear equation for investigating the growth of populations, derived by the Belgian mathematician Pierre-François

Verhulst in 1845, is of the same form as the laser equation. The American mathematician Alfred James Lotka and his Italian colleague Vito Volterra modeled the growth fluctuations of competing prey and predatory fish by coupled nonlinear equations of their populations.

From an ecological standpoint, plant and animal populations are linked to their biochemical environment by a tremendously complex network in which insignificant changes to the existing equilibria (i. e., symmetry breaks) can precipitate natural catastrophes. If the system as a whole attains such critical values but is able to regenerate itself in conformity to a particular macroscopic pattern, the "catastrophe" was a reversible one. But of great interest today is the irreversible symmetry breaking, which is instigated primarily by human intervention in the complex connectivity of nature. Such systems create macroscopic ordered structures spontaneously when their equilibrium states are destabilized by symmetry breaking.

Insect communities, such as those of ants, provide an instructive example. At first sight, they appear to form a deterministic system in which all activities of the individual ants are programmed. But upon closer observation, the individual ants are found to execute many random movements (fluctuations), while the organization as a whole is characterized by highly ordered structures that are, to be sure, liable to change spontaneously. A stable ordered structure may in this case be a network of scent tracks, which the ants construct as they connect their nest to their food sources.

In this situation, the system has attained a state of equilibrium with its environment. But if another, equivalent food source is discovered through the random fluctuations of individual ants, the old network of tracks may become unstable and a new one may be created. In a certain sense, the system will then oscillate between two possible attractors that represent two stable final states, until the system undergoes symmetry breaking and a selection is made in the corresponding bifurcation.

The temporal development of an ant colony is not, as was long believed, completely programmed on the microscopic level of individual ants. On the contrary, a minimal genetic advantage gained by individual ants is sufficient to cause the colony to develop new forms of cooperation on the collective level. The colony is thus capable of solving complex problems when environmental conditions change. Just as in molecular complex systems, one can discern individual, random fluctuations on the microscopic level that may lead to phase transitions and the emergence of new structures on the macroscopic level. The concept of a centrally controlled program or the metaphysical assumption of a higher "controlling time awareness" of the collective consciousness is therefore superfluous. The emergence of collective forms is instead a necessary consequence of the system being far from thermal equilibrium and of the mathematical nonlinearity of its time-dependent evolution equation, while the individuals remain largely autonomous.

Time Hierarchies
and Biological Rhythms

Aside from the irreversible progression from birth to death, the earliest human experiences are surely the repetitive (reversible) cycles such as day and night and the seasons. Humans have used these cycles for their temporal orientation from their beginnings. Since then, researchers working in biochemistry, medicine, and evolutionary biology have discovered diverse temporal rhythms on all organizational levels of life.[8] A network of linked interactions is the mark of a high degree of complexity; these networks exist in organisms on the level of organs, cells, and in the subcellular domain. The biochemical reactions that take place in an organ such as the liver constitute a virtually incalculable number of reaction chains. They are tributaries of a complex metabolic pathway, or else they split off from it to form a new network of interconnected reaction chains. With the aid of the theory of complex dynamic systems and modern computational techniques, it is now possible to analyze complex metabolic processes that involve surprising reactions not previously understood.

Organs must be able to react flexibly to rapid and unexpected changes, in the manner of nonlinear systems. Thus heartbeat or breathing rate must not be tied to a strictly periodic mechanical model such as a pendulum clock. The human body as a whole is itself a complex system in which locally substances are always on the increase or decrease; i. e., birth and creation, dying and decay are taking place continuously. Chaos and self-organization go hand in hand, and only if they are in a state of pre-established harmony do life and health exist.

Against the background of classical mechanics, the heart has for centuries been viewed as a prime example of a biological clock.[9] Aperiodic pathological deviations such as ventricular fibrillation lead to a chaotic final state that means death. A regular heartbeat is contingent on millions of individual heart muscle cells being in phase with each other. Thus a dangerous situation arises if one portion of the heart's muscle cells are in the rest period's nonexcitory state, while another portion is again ready to receive and spread the contraction signal, so that the work done by different regions is no longer coordinated. In particular muscle cell regions, even resting ones, cyclical excitatory processes occur locally that may eventually turn into chaotic ventricular fibrillation. The buildup of electro-physiological potential that is required for doing the contraction work can no longer be achieved, in spite of maximal energy consumption. The pumping action of the heart ceases and circulation comes to a stop.[10]

The vital functions of the organism are maintained by many thousands of reaction cycles, from the molecular biological level to the physiological processes of organs, and each of these cycles has its own characteristic time associated with it. The interior time of an organism is the product of the great temporal rhythms that maintain life, such as heartbeat, hormonal cycle, menstrual cycle, and the wake–sleep cycle on the system's different levels. And the interior time of the organism is itself tied to the great biological and ecological rhythms of nature and of human civilization as well and is dependent on them. There are clinical syndromes that can be traced to disruptions of the interior time linked to various bodily rhythms, such as epileptic attacks, disruptions in breathing rhythms, and heart attacks. Variable psychosomatic disorders are also of this type, as are changes in the complex interconnections in immune or hormonal systems.

The boundary conditions imposed by biological interior times may well call for fresh clinical approaches in the treatment of neurological and mental diseases. Chaos is in the final analysis not a disease state, but a complex attractor. Thus the brain waves measured in electroencephalograms (EEGs) of healthy subjects display chaotically ragged signals, while the EEG curves of epileptic subjects are notably regular and orderly. Lastly, it may be of fundamental importance in therapeutics to gain a fresh and fundamental understanding of the organ under investigation as a complex dynamic system in which sensitivity to miniscule changes must be taken into account.

A hierarchically structured system such as a human being is made up of many different cell types and of complex electro-physiological phenomena, and within it autonomous activities rooted in instabilities take place continuously, from the molecular level to the EEG, from the microcosm to the macrocosm of the brain. A deeper understanding of the temporal phase transitions among the different local conditions might be of considerable value.

1 Andreski, *Herbert Spencer: Structure, Function, and Evolution* (London: Michael Joseph, 1971).

2 Thomson, *Mathematical and Physical Papers* Vol. 5 (Cambridge: Cambridge University Press, 1882–1911), 21–23.

3 Fischer and Mainzer, *Die Frage nach dem Leben* (Munich, Germany: Piper, 1990).

4 Eigen and Schuster, *The Hypercycle: A Principle of Natural Self-Organization* (New York: Springer-Verlag, 1979).

5 Haken, *Synergetics: An Introduction* (New York: Springer-Verlag, 2nd ed. 1978); Mainzer, *Symmetries of Nature: A Handbook for Philosophy of Nature and Science* (New York: Walter De Gruyter, 1996); Mainzer, *Thinking in Complexity: The Complex Dynamics of Matter, Mind, and Mankind* (New York: Springer-Verlag, 3rd ed. 1997); Prigogine, *From Being to Becoming: Time and Complexity in Physical Sciences* (New York: W. H. Freeman, 1981).

6 Prigogine, "Order Through Fluctuation: Self-Organization and Social System," *Evolution and Consciousness: Human Systems in Transition*, eds. Jantsch and Waddington (Boston: Addison Wesley, 1976), 93–126.

7 Turing, "The Chemical Basis of Morphogenesis," *Philosophical Transactions of the Royal Society of London* B 237, no. 641 (1952): 37–72; Gerisch, "Periodische Signale steuern Musterbildung in Zellverbänden," *Naturwissenschaften* 58 (1971): 430–38.

8 Hess and Markus, "Chemische Uhren," *Selbstorganisation: Die Entstehung von Ordnung in Natur und Gesellschaft*, eds. Dress, Endrichs, and Küppers (Munich, Germany: Piper, 1986), 61–79.

9 Mainzer, "Chaos und Selbstorganisation als medizinische Paradigmen," *Wissenschaftstheorien in der Medizin*, eds. Deppert, Kliemt, Lohff, and Schaefer (Berlin, Germany: Walter de Gruyter: 1992), 225–58.

10 Winfree, *When Time Breaks Down: The Three-Dimensional Dynamics of Electrochemical Waves and Cardiac Arrhythmias* (Princeton, New Jersey: Princeton University Press: 1987).

chapter 7

Time and Consciousness

Different temporal rhythms may be distinguished among the coordinated physiological processes that take place in the brain. The phenomenon of time consciousness is intimately connected with the dynamics of the brain's different states of consciousness. The theory of complex systems far from thermal equilibrium has been used to formulate explanations for the emergence of consciousness. Consciousness is acordingly understood as a macroscopic, global-ordered state of neural switching patterns that result from microscopic local interactions in the complex networks of the brain. Time consciousness is not contradicted by physics but it can be explained as resulting from a complex neural interaction process. From the point of view of computer science, the question arises about what connections exist between computer machine time and human time consciousness.

Temporal Rhythms
and Brain Physiology

In the framework of complex systems, the brain is an assemblage of approximately 10 to 100 billion interacting neurons. Its superiority over computers developed to date does not lie in its information processing speed or the precision with which it operates. In contrast to the early, high-speed computer developed by John von Neumann, a large portion of the brain's neuron population is active simultaneously, with different portions communicating with each other. These complex networks are prerequisites for our ability to coordinate activities such as movement, recognizing people, or conversing with others in just fractions of a second. When asked to perform similar tasks, programmatically controlled computers fail significantly.

From the point of view of systems theory, it would seem appropriate to describe the temporal evolution of neurons and their interactions using equations that are analogous to those used for cell populations in organisms (see page 114). However in the case of the brain, one would have to take into account the millions of temporal developments of the individual neurons as well as many additional parameters. For this reason, it becomes evident that a true simulation of the brain or a numerical solution for its activities is quite impossible. Instead, the structural principles of the brain need to be discovered and described mathematically in order to establish the basis for understanding the emergence of thinking, feeling, and speaking.[1]

A crucial capability acquired by the brain in the course of its phylogenic evolution is the capacity of some neurons to make autonomous synaptic connections. Such spontaneous modifications of synaptic connections are responsible for the brain's learning ability. These may be modeled by assigning variable weights to synapses, since the effectiveness of interneuron connections (associations) depends on the corresponding synapses. From the physiological point of view, learning presents itself as a local temporal process. The modifications that occur in the synapses in time are not controlled from outside, but occur locally in individual neurons through processes such as the modification of the amount of neurotransmitters or the growth of new synaptic connections.

This time-dependent modification of the synaptic weightings takes place according to a learning rule. Donald Olding Hebb, a cognitive psychobiologist, suggested that such a rule might involve the strengthening the connections between neurons that are frequently activated together.[2] This leads to the generation of activity patterns (assemblies) and the formation of neural correlations in the brain that correspond, in turn, to signals from the outside world. Such neural patterns may correspond to words, sounds, images, or entire situations.

With this model, several capabilities of the brain can be understood. If one wishes to commit something to memory, the corresponding activity pattern is retained in the brain by repeatedly activating the content it represents. According to Hebbs's rule, this repetition reinforces the connections between the activated neurons. If one wishes to recall something, the entire content matter must be reconstructed from a portion of it. Hebbs considered this kind of pattern completion to occur spontaneously when some of the neurons in a learned pattern are activated. The activity patterns may also correspond to abstract concepts such as geometrical shapes. The learning process creates connections between neurons and groups of neurons in a network, so that a thought association, once it has been learned, is a kind of temporally "condensed" correlation pattern.

The Experience of Time and the Emergence of Consciousness

In order to understand the experience of time and time awareness, one must first understand the emergence of consciousness in biological evolution. A remarkable hint relevant to this discussion can be found in the writings of Gottfried Wilhelm Leibniz, the famous German philosopher and mathematician.[3] Leibniz argued that if we visualize the brain as a mechanical clock, that has been greatly enlarged so that we can enter it like a mill, we would discover interacting toothed gears, not thoughts and feelings. Thoughts and feelings are thus to be understood as functions of the entire mechanism, just as a clock's individual mechanical components and their local interactions collectively indicate the clock time.

However, the brain is not an inanimate mechanical clock but rather a living system far from thermal equilibrium. In terms of the theory of complex systems, the brain may be considered as a complex aggregate of nerve cells that form networks as a result of phase transitions and create new patterns through self-organization. The macroscopic switching pattern might correspond to external perceptions, emotional states, or thoughts, as well as an awareness of time. Perception is not to be regarded as a rigid and isomorphic image of the external world, but as a learning process, in which an image of the outside world is created step by step and is subject to constant correction.[4]

Brain physiologists have already investigated and made maps of sensory and motor neuro-activity as representations of perceptions and movements and their sensory coordination. Investigations of the neural activity patterns evoked by emotions and thoughts are still in their infancy, but the theory of complex dynamic systems has been applied to problems in many disciplines, including physics, chemistry, and biology and provides heuristic hints for research hypotheses in these investigations.[5]

To understand the phenomenon of time awareness, an explanation of consciousness is necessary. Brain researchers and cognitive psychologists have suggested explanations that can be modeled within a theory of complex dynamic systems.[6] First, we must remember that the perceived content of the external world can be represented by a typical neural switching pattern. If one thinks about this perceived content—an act traditionally referred to as reflection—a new thought is created that makes reference to a preceding one. One can now imagine that the output of the cerebral switching pattern that represents the preceding thought serves as the input for the neural pattern switched on by the subsequent thought and initiates self-organization processes in it.

This generates a meta-representation that can be iterated at will: I am thinking about how I was thinking about thinking about . . . thinking. The iterated meta-representations of neural switching patterns are supposed to model the formation of the state of consciousness. The degree of consciousness depends on the speed with which the iterated meta-representations can be formed. The switching speed of the synapses can be affected chemically by means of therapeutic and other drugs, allowing changes in consciousness to be tested experimentally.

The phenomenon of time awareness can be considered in a like manner. Historically, time has been represented as "pure duration." According to Henri Bergson (1859–1941), time was a pure intuition, to be understood as a human being's look into his inner self. In other words, intuitively "experienced" time was a continuous duration and thus not divisible. Bergson criticized classical physics' concept of time, which represented time geometrically as a straight line with time points and duration measured as a distance in time. In classical mechanics, relativity theory, and quantum mechanics, time is employed as a real mathematical parameter, and the laws of these physical theories are time symmetrical, just as a straight line has mirror symmetry with respect to the origin.

In subjectively experienced time, the present may last an eternity, may flow sluggishly, or may elapse instantaneously. To be sure, relativity theory states that every reference frame possesses its own proper time (*eigenzeit*), a mathematically precise and objectively quantifiable concept. Subjectively experienced time is more than this. In the end, Bergson said that, in the intuition of duration, we experience our own person in passage through time: "It is our ego, which endures in time."[7] Bergson maintained that, in this intuition of time duration, a multitude of successive states of consciousness can be distinguished abstractly, while we also recognize a unity binding them together.

Bergson would surely agree that time awareness, just like other perceptions, can only be experienced subjectively. Nobody has echoed these sentiments better than his compatriot Marcel Proust, who explored the subjective proper times of specific life processes and the loss of a global time of shared experience in his novel *A la recherche du temps perdu.* The dynamics of time awareness can nevertheless be subjected to scientific observation and analysis, just as we study the dynamics of other perceptions. Our studies are inherently limited, however, because we cannot directly experience someone else's subjective time sense. Fortunately, the goal is to gain a better understanding of the physiological and psychological conditions related to these states and draw conclusions that may be of value in medical, psychiatric, or psychological diagnoses or therapies.

In this manner, Ernst Pöppel characterizes consciousness processes in information processing systems whose functions are synchronized in time.[8] By integrating over time, it is possible to sum up the events up to a certain limit as perception and awareness structures. The integration time is limited to nearly 3 seconds, so that the subjectively experienced present, i. e., awareness of the Now, corresponds at most to that interval. Timing errors caused by over- or underestimation of an "instant," are on the order of two or three seconds and are interpreted as the temporal limit of the brain's integrating power.

In accordance with this approach, "consciousness" is defined as a state, in which several sensorial stimuli and mental functions are integrated for a few seconds, standing at the focus of one's attention. According to this analysis, the awareness of a continuous current of time is an illusion that results from stringing together successive states of consciousness of about three seconds' duration each. Consciousness, whether self-consciousness or time awareness, is not a fundamental control unit, nor an ontological substance, nor a mysterious stream of experience that cannot be further analyzed. Instead, it signifies a macroscopic state that is generated by neural networks under particular conditions in a process of dissipative self-organization.

Computation Time and Artificial Intelligence

Computation time is a measure of the effort and the complexity needed to solve a problem by machine. For this purpose, a computational problem is represented by a function that can be calculated by a computer ("Turing machine") in a finite number of basic steps (e. g., changing a 0 to a 1). These computational steps serve as time units that are counted in the process of performing a computation.[9]

As a measure of a problem's complexity, one focuses on the running time and data storage requirements of an algorithm and their dependence on the length of the input. The running time of an algorithm is given by the number of steps taken in executing it when presented with an input, while the storage requirements are quantified by the number of requisite storage cells.

In considering an algorithm's complexity, one distinguishes between the worst-case and average-case complexities, with the former corresponding to the longest running time and greatest storage requirement for an input of a specific length. If some inputs occur more frequently than others, the average-case complexity is obtained by weighting the inputs according to a probability distribution.

In general, complexity theory may be defined as the discipline that deals with the complexity of algorithms and functions. Apart from the above-mentioned running time and storage requirements, the theory deals with the classification of problems into complexity classes, according to the dependence of running time on input length.

If the running time for all inputs of length n is proportional to n, one speaks of a linear running time; if it is proportional to n^2, the running time is said to be quadratic. Similarly, the running time is said to be polynomial if it is proportional to a polynomial $p(n)$, and it is exponential if it is proportional to an exponential function $2^{p(n)}$.

One must also distinguish between computational times for machines that operate deterministically versus those that operate nondeterministically. While the sequence of commands is uniquely determined in a deterministic machine, a nondeterministic machine contains instructions for nondeterministically executing one of a finite set of instructions. An algorithm is said to be polynomially time-limited if it can be calculated in a polynomial running time by a deterministic Turing machine. The class of all functions that are calculable by a polynomially time-limited algorithm is denoted by P, and the class of all functions that can be computed by a nondeterministic Turing machine in a polynomial running time is denoted by the symbol NP.

A fundamental problem of complexity theory is whether or not the class of problems computable in polynomial time by nondeterministic Turing machines is the same as the class of problems computable in polynomial time by deterministic Turing machines. In short, is the statement $P = NP$ valid? As it turns out, computational time is a fundamental measure of problem solving by machine.

At first, scientists suspected that appreciably shorter computational times were achievable with computers operating on the basis of quantum mechanics and not according to the principles of classical physics.[10] After all, quantum field theories describe the dynamics of elementary particles (e. g., photons) that travel at exceedingly high speeds. A quantum computer is a generalization of a classical computing machine (a Turing machine), comparable to a quantum-mechanical generalization of a classical physical system, according to the Bohr correspondence principle (see the first section of Chapter 4). A classical Turing machine thus represents a special case of a quantum computer whose dynamic properties ensure that the state of a calculation be read out after each computation step. In a quantum computer, this would in general be impossible, since the "act of reading out" would, in the sense of the quantum-mechanical measurement problem, change the computational states of the machine. But in contrast to a classical machine, the quantum computer also accepts additional programs that transform the computational states into linear superpositions that are nonseparable in the sense of quantum-mechanical nonlocalization.

Such superpositions of states suggest an analogy to parallel computing techniques performed with classical machines. A quantum computer might prove useful if one is interested in a particular correlation of a very large number of computational results, but not in their individual details. In the latter case, a quantum computer could superimpose a myriad of parallel computations with an efficiency that surpasses classical computers. Instead of speaking of a superposition of computational states, the term *quantum parallelism* is used, in contrast to the classical parallelism of conventional Turing-type computers.

It can, however, be proved that the expectation value of the computational time for quantum parallelism cannot be shorter than for a serial computation by a quantum computer. Furthermore, quantum computers operate algorith-mically just as before, for their linear dynamics is a deter-ministic process. The nondeterministic aspect surfaces only in the nonlinear act of measurement in the sense of the quantum-mechanical measurement problem. It follows that it cannot be expected that a quantum computer can perform nonalgorithmic operations beyond the complexity limit of a Turing machine.

In the framework of complexity theory, a quantum computer may nevertheless possess technical and practical advantages over conventional classical computers. In princi-ple, one can imagine quantum computers operating at higher speeds and solving problems in polynomial time, although they do not belong to the complexity class P.

As classical computers are based on classical physics, and quantum computers on quantum mechanics, both kinds of computers are based on the concept of time reversibility. The laws of nature under which they operate permit, in principle, their computing processes (other than the act of measurement and reading out in the case of quantum computers) to run backward in time. This raises the ques-tion of whether it might also be possible to use computers to simulate time-irreversible processes that are well known from biological evolution and the self-organization of the brain. The emergence of cellular patterns was simulated for the first time in the 1950s by von Neumann's "cellular automata," and many computer experiments since then show the emergence of patterns that are familiar as the attractors of complex dynamic systems.[11] One encounters classes of patterns that tend toward an equilibrium state ("fixed point"), others that tend toward periodic repetitions ("limit cycles" and "oscillations"), as well as fractal chaotic patterns. The latter, in particular, display the typical sensitivity of chaotic systems to initial conditions (the *butterfly effect*). Consequently, they represent developments of irreversible patterns as in biological evolution, whose course would be altered by the most minute changes in the initial conditions. Because of this sensitivity, it is impossible to predict such complex developments in detail.

An early attempt to use a computer to simulate brain function made use of the so-called *McCulloch-Pitts networks*. In 1943, Warren S. McCulloch and Walter Pitts described a complex model of artificial neurons equipped with excitory and inhibitory synapses, which served as threshold logic units. A McCulloch-Pitts network consists of interconnected McCulloch-Pitts neurons with the following properties: When at time t the weighted sum of inputs received from other neurons exceeds a certain threshold, the neuron fires an output impulse along its axon at time $t + 1$. The output pulse of each neuron is split and channeled to some of the other neurons as input pulses. Thus, each neuron serves as a logic gate, so that a Turing machine can be used to simulate the entire McCulloch-Pitts network. In this model, the weights and connections of all "neurons" are, however, fixed for all times. This eliminates a critical capability—developed in the course of the brain's phylogenic evolution—for the modification of interneuronic synapses, which makes the learning process possible. Learning is nothing other than a special form of self-organization that can take place irreversibly in dissipative ("hot") systems far from thermal equilibrium. The massive parallel processing of information is also characteristic of the brain. In contrast to traditional, serially processing computers, we can simultaneously perceive, feel, and think a number of different things.

The paradigms of parallelism and connectivity are of current interest to engineers engaged in the design of neurocomputers and neural networks.[12] In principle, it is not impossible that this approach will result in a technically feasible neural self-organization that leads to systems with consciousness, and specifically with time awareness. On the other hand, such systems would lack our human awareness of time, which depends on the internal time of our biological and socio-cultural evolution. Models of this kind are merely useful for investigating rules governing mental processes and their dependence on neural events in the brain, but are not useful for studying individual thoughts or intimate feelings. It is, after all, characteristic of complex dynamic systems that their temporal evolution cannot be predicted in detail, although they may display typical developmental patterns under critical boundary conditions.

1 Mainzer, *Computer, neue Flügel des Geistes? Die Evolution computergestützter Technik, Wissenschaft, Kultur und Philosophie* (Berlin, Germany: Walter de Gruyter, 1994).

2 Hebb, *The Organization of Behavior: A Neuropsychological Theory* (New York: John Wiley & Sons, 1949).

3 Leibniz, *Monadology* § 17, ed. Rescher (Pittsburgh: University of Pittsburgh Press, 1991).

4 Kohonen, *Self-Organization and Associative Memory* (New York: Springer-Verlag, 1989).

5 Ciompi, "Die Hypothese der Affektlogik," *Spektrum der Wissenschaft* (February 1993): 76–87.

6 Flohr, "Brain Processes and Phenomenal Consciousness: A New and Specific Hypothesis," *Theory and Psychology* 1 (1991): 245–62.

7 Bergson, *An Introduction to Metaphysics* (Kila, Montana: Kessinger Publishing, reprint ed. 1998).

8 Pöppel, "Eine neurophysiologische Definition des Zustands 'bewußt'," *Gehirn und Bewußtsein* (Weinheim, Germany: Wiley-VCH, 1989), 17–32.

9 Traub, ed., *Algorithms and Complexity: Recent Results and New Directions, Proceedings of a Symposium on New Directions and Recent Results in Algorithms and Complexity held by the Computer Science Department, Carnegie-Mellon University, April 7–9, 1976* (New York: Academic Press, 1976).

10 Deutsch, "Quantum Theory, the Church-Turing Principles, and the Universal Quantum Computer," *Proceedings of the Royal Society of London* A 400 (1985): 97–117.

11 Farmer, *Cellular Automata* (New York: Elsevier Science, 1984).

12 Eckmiller, *Parallel Processing in Neural Systems and Computers* (Amsterdam, Netherlands: Elsevier/North-Holland Publishing Co., 1990).

chapter 8

Time in History
and Culture

Historical cultures, like individuals, developed different internal times in the course of their evolution. As a result, philosophers of history have offered different temporal models to explain the birth and demise of historical cultures. The theory of complex system also allows us to model the dynamical development of social, economic, and cultural systems. In this chapter, we will learn that at least some aspects of irreversible temporal developments in human society may be analyzed by methods analogous to those used for physical and biological processes. But this does not imply a naturalistic reductionism. In historical and technological cultures, time represents the emergence of a new phase of biological and socio-cultural evolution.

Time in Historical Cultures

People from around the world and throughout time have had a strong sense of the history of their groups, tribes, nations, and cultures. These collective temporal memories were originally transmitted orally in myths and stories. Eventually, some groups began writing the stories down. Beginning in the era of the ancient high cultures, these writings were developed further by historians like Herodotus and Thucydides. In the Aristotelian tradition, history strictly involved the presentation of individual facts in time. A philosophy of history, on the other hand, explores the causal relationships between historical developments. Saint Augustine, one of the earliest philosophers of history, interpreted history as a struggle between God's realm (*civitas Dei*) and the godless powers (*civitas diaboli*). Later, medieval historical theology already distinguished several epochs, in which the temporal development proceeds eschatologically from Creation toward the Last Judgment.

In the modern philosophy of history, Giambattista Vico plays a central role. In *Principi di una Scienza Nuova* (1725), he interprets history as a sequence of eras of cultural growth and decay. In the higher stages of civilization, Vico believed that there were repetitions of lifestyle and conduct as new conditions develop. For example, Vico saw a connection between the heroic barbarism of bygone eras and the neo-barbarism of more highly developed societies. Since Vico, other philosophers have attempted to describe historical temporal evolution in terms of the laws of development and mechanics.

In the age of Enlightenment, history was seen as the progress of reason and the gradual abandonment of superstition. In his 1784 treatise "Idea for a Universal History with a Cosmopolitan Intent," Immanuel Kant explains history as progress toward a perfect civic unification of all people. Kant describes this ultimate state of human history as not just a pleasing fantasy, but as a possibility based on human nature. People's "asocial sociability," an anthropological term used by Kant that refers to our tendency to seek out social contact while also seeking to expand and defend our territorial boundaries, demands regulation guided by reason. Only through such regulation can we hope for the peaceful coexistence of all people on this Earth. This coexistence, according to Kant, is what constitutes the cosmopolitan society.

Approaching the question from a slightly different angle, Georg Wilhelm Friedrich Hegel argued that the history of the world proceeds in dialectical steps as the self-realization of the spirit toward ever-greater freedom. For Hegel, the climax of this temporal development would be attained in the condition of a civil society, which is distinguished by its legally and constitutionally founded institutions. In the course of this historical process, both the subjective consciousness and the personal will of individual human beings are overcome and preserved in the collective ("objective") consciousness with a general and free will that is realized in the civil state.

Karl Marx interprets Hegel's law of dialectical development as economic class struggle, which leads to the classless society as its final state. Auguste Comte, on the other hand, describes history as the linear progression of three stages. According to Comte, humanity passes first through a theological stage, dominated by religion and myth, then moves into a metaphysical stage, in which the world is perceived in philosophical and metaphysical terms, and finally enters a scientific stage dominated by technology and science.

In addition to Hegel's and Comte's laws for the historical development of human society and culture, nineteenth-century natural scientists also began formulating laws for the temporal development of nature—in evolutionary biology and in thermodynamics. These ideas inspired Herbert Spencer to consider the socio-cultural evolution of human society as a linear continuation of the biological evolution of life, which is embedded in the cosmic evolution of the Universe. According to Spencer, evolutionary progress can be measured in terms of growing complexity. Evolution proceeds, he says, from a state of incoherent homogeneity to states of coherent heterogeneity.

Although many comments of Spencer's anticipate the modern theory of complex dynamic systems, he is today regarded as a naturalist and reductionist thinker caught up in the nineteenth century's linear concepts of time and social evolution. Thus, while he saw the temporal developments of cosmic, biological, and societal evolution as irreversible processes, they are dealt with in terms of equilibrium thermodynamics. Spencer considered the emergence of new structures and organizations in society to be based on the same principles as the appearance of a crystalline frost flower under thermal equilibrium conditions.

When the idea of progress became discredited in the wake of political and cultural changes at the beginning of the twentieth century, time-based models again gained the upper hand. These models either leaned toward Nietzsche's nihilistic belief in a meaningless cycle of history (*amor fati*) or reached back to the organic imagery of the growth and decay of cultures. Thus, with a glance toward Nietzsche and von Goethe, Oswald Spengler proposes a "morphology" of world history, in which the metamorphosis of Western culture, as it passes from its first budding ("spring") through a mature stage ("summer") and harvest ("autumn") to decadence ("winter"), is compared to corresponding developments in Indian, classical, and Arab cultures. Cultures accordingly resemble independent organisms, with their own rhythms in time and life, and like organisms they have a beginning and an end in time.

Arnold Toynbee does not follow Spengler's cultural morphology. Instead, his account of the cultural development of humanity is guided by Bergson's temporal philosophy. Toynbee considers an archetypical culture to be a fiction. Instead, the development of all populations is continuous but distinct, each proceeding according to the particular prevailing historical circumstances and without displaying the schematic commonalities envisioned by Spengler. According to Toynbee, cultures, like organisms, are able to combine and influence each other. While Spengler assumes the existence of 8 cultural archetypes, Toynbee distinguishes 21 cultures, each with its own characteristic rhythms of temporal development.

Another attempt at a kind of historical morphology was undertaken by Karl Jaspers in his book *The Origin and Goal of History* (1949).[1] Jaspers claims to have discovered an "axial time," or the axis of world history, which commences in approximately 500 B.C. He believes that the common spiritual foundations of the world's different cultures—which made the emergence of modern humanity possible—was established between 800 and 200 B.C. This era is pivotal because of Confucius and Lao Tzu living and teaching in China, Buddha in India, Zoroaster in Persia, the prophets in Palestine, and the classic philosophers in Greece.

In that epoch, all attitudes, customs, and conditions, until then accepted unconsciously, were put to the test and questioned, suspended, or newly formulated. In analogy to the new spiritual world, a new political and sociological state emerged. The ancient high cultures, isolated from each other for thousands of years, lost their significance and common spiritual foundations when ethics, science, and culture were created. Jaspers believes that beginning in this axial time, world history acquired the common structure and unity that persist to this day.

Jaspers also speaks generally of the historicity of human existence, an idea that describes our existence as embedded in humanity's cultural, social, economic, political, and world history. In discussing the history of human cultures, Martin Heidegger puts great emphasis on temporality as a fundamental structure of human existence, which stretches from birth to death. Comparisons and distinctions between individual fates and cultures are only possible because of the common time horizon of human existence: existence is, in the famous words of Heidegger, "being-to-death."

Karl Popper, among others, raises crucial objections to a fundamental ontology of temporality and to interpretations or explanations of history that are based on speculative universal laws. Pronouncements derived from them are, according to Popper, arbitrary and not binding, since their methodology is ambiguous and cannot be tested. The neo-Kantian tradition had already stressed the special status of statements about time in the context of the philosophy of history. Wilhelm Windelband maintains that contrary to the law-writing (nomothetic) natural sciences, the writing of history is, like all humanities, idiographic; i. e., it explores the details of unique and irreproducible events.

Georg Simmel and Wilhelm Dilthey conclude that only understanding and empathy for these oddities of historical writing can provide an adequate methodology. This position comes close to the guiding philosophies of Leopold von Ranke and Johann Gustav Droysen, both of whom were skeptical about speculative "laws of motion" for history such as Hegel's.[2] Popper considers these general pronouncements about the time course of history to be sheer prophecies, independent of prevailing boundary conditions. He maintains that they must be distinguished from prognoses based on the hypothetical-empirical laws of the social sciences. Max Weber had already pointed out that researchers who study history and social sciences aspire to use objectively controllable methods of investigation, so that temporal and historical events can be related to general archetypes, which Weber called "*Idealtypen*."[3]

In analytical philosophy and the philosophy of science, scholars like Carl Gustav Hempel and Ernest Nagel, among others, have explored a general methodology for explanations within the natural, social, and historical sciences, in so far as these shall be considered empirical sciences.[4] They contend that the neo-Kantian distinction between nomothetic and idiographic disciplines is obsolete, because the natural sciences also investigate unique, irreproducible events and processes such as biological evolution that are irreversible in the thermodynamic sense. Within the framework of the theory of complex dynamic systems, scholars from different disciplines are attempting to determine the subsidiary conditions under which the emergence of new structures, organisms, organizations, etc., can be explained according to particular laws. As a result of these queries, the question arises, Which aspects of the socio-cultural evolution lend themselves to this kind of mathematical modeling?

Time in Technological-Industrial Cultures

How can the emergence in time of political, social, and economic structures in human society be explained? Historically, the concepts of time derived from nature or from technology were used to seek explanations. Early in the modern age, Thomas Hobbes applied the Galilean and Cartesian laws of motion of mechanics to anthropology and political science. In his "leviathan model," the state is organized like a centrally controlled machine, in which the citizens interact in their respective functions like the machine's gears. The physiocrats, a group of French economists, compared the economic system of the autocratic state of the French monarchy to an eighteenth-century mechanical clock. In this model the economic interactions between landowners (the nobility, the Church), farmers, artisans, and merchants are strictly regulated. The passage of economic time is likened to balls rolling along the fixed tracks of a working clock of a type used at that time.[5]

The liberal ideas about the state and economy espoused by John Locke, David Hume, and Adam Smith were developed against the background of the time concepts of Newtonian physics. Contrary to the Cartesian mechanics of levers and gears that inspired the physiocrats, Newton's gravitational theory envisions forces acting over a distance, forces that cause freely floating celestial bodies to interact and develop a sustained state of equilibrium. Similarly, Adam Smith argued that just as gravity acts invisibly in physics, so will an "invisible hand" establish the market equilibrium between supply and demand, leading to a "natural price." To arrive at this conclusion, Smith presupposed that—because of their nature—all economic agents will seek to maximize their profits.

Against the background of nineteenth-century classical physics, the Lausanne school of Léon Walras and Vilfredo

Pareto promoted the study of linear dynamic systems in economics. Their concepts of time were also borrowed from classical mechanics and equilibrium thermodynamics. They employed technical terms such as equilibrium, stability, elasticity, expansion, time course, resistance, and friction. The actions of individual humans (*Homo oeconomicus*) were presumed to be rational and predictable in time.

However, the use of linear models for the overall temporal development of society precludes the emergence of any nonlinearities, chaos, or synergistic effects. The classical economists only considered linear equilibrium systems because their equations were easier to solve than nonlinear ones. Just like Laplace in his celestial calculations ("Laplacian spirit"), the followers of the Lausanne school assumed the existence of an economic reality with a calculable, long-term development in time. But since the market economy is an open system that is constantly exchanging materials and energy with other markets and with nature, it cannot be an equilibrium system. Economic systems, just like biological systems, are constantly changing and are very sensitive to minute changes in their boundary conditions. Short-term fluctuations in consumer preferences, inflexible reactions to changes in productivity, as well as speculations in raw materials and real estate markets serve to illustrate the sensitive reactions of economic systems. The centuries since Adam Smith outlined his economic theories have shown that fluctuations on a small scale may—through self-organization—lead to large-scale bursts in growth (e. g., technical innovations such as mechanical looms or the steam engine in the industrial revolution). On the other hand, they may also escalate into chaotic, uncontrollable behavior (e. g., stock market crashes).[6]

It follows that cyclical behavior and chaos are not exceptions, but a part of economic reality. The temporal developments in a complex dynamic system are conveniently represented by trajectories that may head for different attractors. From the point of view of complex systems theory, the temporal development of human society is explained in terms of evolving macroscopic order parameters (e. g., economic or social ordered states) that originate on the microscopic level from nonlinear interactions between individuals and subgroups like firms, institutions, and nations.

These developments are clearly revealed in economic time series analyses.[7] Chaotic market fluctuations, such as those of the Great Depression at the end of the 1920s, were originally investigated by means of linear models. In order to understand irregular departures from these models, it was necessary to assume *ad hoc* that exogenous shocks occurred, without however being able justify them in economic terms. This is equivalent to explaining the irregular vibrations of a string by saying they are caused by external forces. It was not appreciated at the time that market cycles can also have endogenous causes and that these cycles are intrinsic properties of an economic system. On the other hand, nonlinear systems in which the temporal development of different economic parameters are coupled nonlinearly to each other, admit endogenous, chaotically irregular temporal developments, whose long-term fates are as difficult to calculate as the planetary orbits in Poincaré's famous three-body problem (see page 98).

Methodological problems in economic time series analyses arise when we attempt to design of empirical tests and determine their efficacy. This is so even though incomparably more and higher-quality quantitative and observational data is available today. In contrast to the natural sciences, where measurements are often arbitrarily precise and experimental checks are feasible, time series analyses frequently remain constrained by coarse units of time such as days, months, or years. The standard time series typically has a length of only a few hundred units, so that any economic models based on it have limited reliability on empirical grounds alone. Empirical experiments are, of course, largely impossible to carry out.

Nonlinear time series analysis is well suited for handling the structural changes in economic growth that have been initiated by the expansion of the new high-tech industries. Traditional economic theory presupposes that the supply of natural resources for industries is always decreasing. The more of these raw materials that are produced and reach the market, the more problematic mining and production becomes, causing profits to decrease. Consequently, the temporal development of these classical industries is controlled by a negative feedback mechanism. This is in contrast to the new growth industries like electronics and computer and information technologies, which are essentially independent of natural resources. They depend instead on increasing knowledge, and hence display novel positive amplification and feedback effects.[8]

On the operational level, a time series analysis also provides astonishing insights into the changes that have taken place in our technological-industrial world. A concept of time based on deterministic algorithms along the lines of a punch clock is only suitable for the case of the semi-skilled assembly-line worker who executes a single operation at a prescribed pace during the manufacture of standardized products (e. g., in the automobile industry à la Henry Ford). In the operational situations of today's complex knowledge-based systems, each worker is assumed to be capable of exercising a certain amount of judgment. Thus, individual work schedules and coordination lead to new ordered structures that are appropriate for particular situations and differ from the command structure of a traditional bureaucracy. From the standpoint of complex dynamic systems, this represents the beginning of dissipative self-organization.

The long-term development of cities also lends itself to an analysis as a complex system undergoing phase transitions. It becomes apparent that cities develop not as the result of Cartesian planning, but as a result of long-term self-organization processes. Thus, in a city considered as a complex social system, certain regional population trends are assumed to occur on the microscopic level, where nonlinear interactions (themselves functions of space, traffic, recreation, economics, etc.) lead to changing patterns of settlement on the macroscopic level.

Cities, institutions, corporations, and administrations can develop their own temporal rhythms that are analogous to the rhythms of biological organisms. Brasilia is a rare example of a city planned along Cartesian principles; it was constructed all at once in a uniform style, like a technical system. Rome, on the other hand, is the collective product of different epochs of history and style, each with its diverse rhythms of development, living side by side. As a result, it seems possible to discern the interior times of cities, states, and political and economic institutions. They are new examples of interior times of more complex dynamical systems that cannot be reduced to the measures of time in the less complex systems of physics and biology, or the psychological awareness of time in individuals.

It follows that, we can no longer speak of a universal measure of time in the domains of politics and economics. Instead, different political components, such as parliaments, administrations, and governments, develop their own characteristic time periods that are themselves overlaid with individual conceptions of time (e. g., practicing politicians) and the temporal rhythms of the environment (e. g., cyclical natural events or economic cycles). Specifically, the plebiscite-based, representative, parliamentary, and presidential systems in a democracy have each developed their own time structures and the corresponding differences have been related to the historical growth rates of governmental systems in countries like in France, the UK, the US, and Switzerland. The practical consequences are unmistakable, as is evident in the process of European unification.

In a democracy, the very essence of politics is the control of time, time that is structured by legislative sessions, terms of office, election periods, periods of government formation, and daily agendas. In ancient cultures and pre-industrial societies, the determination of time was also used to validate political authority, but the determination of time is itself rooted in the concept of a universal time (e. g., astronomical time in Egypt and Babylonia, the timetable of Creation in the Middle Ages). Historically, calendar reform has been used to achieve dominance, as is seen in the actions of the Roman Caesars and popes and the French and Russian revolutionaries.

There is a striking parallel between the political analysis of time and the conception of time in physics. From the seventeenth to the nineteenth century, Newton's classical physics assumed the existence of an absolute time of the Universe, to which all clocks could in principle be synchronized. Newtonian absolute time was also God's time, a medium through which He demonstrated the power of His creation. After Einstein, only the relative proper times ("*eigenzeit*") of different physical reference systems were deemed to exist, and it was concluded that no absolute synchronicity can be established on account of the finite speed of all signal transmissions.

The use of the term *eigenzeit* in reference to political systems and institutions reflects an analogy between the terminology of political science and sociology and that of relativity theory.[9] In view of the slow pace of the political systems of this Earth, their physical proper times in the sense of relativity theory are hardly significant. In the present context, proper time instead refers to a system's interior time, which is revealed in the different irreversible phase transitions marking its evolution. In any case, it should be clear that references to the proper time, or the interior time of political or social systems, do not imply a postmodern relativism, but are clearly defined in the particular scientific discipline.

The Time Horizon of the Technological World and the Philosophy of Time

Socio-cultural evolution has brought us to the present technical-scientific world. What is the appropriate time horizon for this emergent science-based world civilization? Today, the information network that spans the whole Earth has made it possible for the individual and his or her consciousness to be a part of a global communication medium. Will the consciousness of the individual dissolve into a sort of world brain with a common time awareness, as Marshall McLuhan prophesied at the end of the 1970s? The combination of computer and communications technology has already produced an electronic infrastructure with far-reaching social and economic consequences. Today's much-discussed transition from the industrial society, built predominantly on material resources, to the information society, in which immaterial values like information and time become scarce goods for providers and consumers, is a consequence of technological innovation. This development is by no means centrally controlled: instead, ordered structures emerge, virtually through self-organization, out of the apparent chaos of the information carriers.

It is the goal of the CSCW ("computer-supported cooperative work") concept to bring people at different locations together in a virtual space at the same time, so that they can deal with the same documents and data as if they were meeting in person. In the long run, two-dimensional meetings via video screens may be replaced by the three-dimensional representations in space and time (e. g., by holographic methods) to facilitate communication over intercontinental distances at any time. This makes it technically feasible to establish a common time horizon of humanity's world society.

Some people see this as a vision of boundless and instantaneous understanding, but others perceive such gigantic computer networks as a threat, as in Hobbes's *Leviathan*. The digitization of all forms of human expression and the global access to information and people could also facilitate control measures that states or monopolistic industries might use to secretly undermine civil rights and freedoms.

Computer and information networks evidently share some of their characteristics with social and biological organizations.[10] They are open complex systems with nonlinear interactions that steer them toward different equilibrium situations ranging from homogeneous final states and oscillations between states to information chaos. The competition among information entities appears to be determined by self-organizing market mechanisms that are reminiscent of economic systems. Based on the rate at which chance biochemical mutations occur, biological evolution takes place comparatively slowly, while changes in humanity's technological cultures can be launched very quickly by unforeseen ideas and innovations.

Thus, while biological systems employ genes as their replicators and mutations as the variational mechanisms, economic systems are composed of individual firms, innovators, and market mechanisms. The replicators of human cultures are information patterns that in the early stages were transmitted by imitation from person to person and from generation to generation. In an analogy to biological genes, these patterns are referred to as *memes*, and they may encompass ideas, beliefs, opinions, manners, fashions, or techniques.

These memes evidently have proper times that differ from those of the human beings who formed them with their technology and culture and who reflect them in their consciousness. While the biological evolution of the past ten thousand years has had almost no effect on the human gene pool, the universe of human thought, action, and feeling has been profoundly altered by changes in culture and technology. There are already speculations that the familiar, genetically directed evolution has reached its limits in the case of human beings, and that it will be replaced by a "post-biological" era of technological culture, which generates its own replicators of information and heritage.

The reader is reminded of the developing technology for creating a cyberspace, in which users see and perceive their own bodies in real time in a three-dimensional virtual space, to the extent that it is technically feasible for their sensory organs to be simulated. The American robot specialist Hans Moravec foresees an even more dramatic development in the form of a "bioadaptor," which will establish a direct connection between brain and computer.[11] Here, Moravec assumes that, in this way, streams of information could be transferred to a machine, creating a complete duplicate of a flesh-and-blood person's ego. Biological cloning seems outright archaic when compared to such visionary computer technologies. Science fiction visions become conceivable: astronauts might leave their actual bodies behind and head for distant stars. Concepts such as life span, death, and personal identity would lose their present meaning if it became possible to separate, transfer, copy, and alter consciousness, intelligence, and feeling from the persons possessing them.

Moravec's computer technology speculations about novel forms of time awareness and consciousness converge with the cosmological scenarios envisioned at the end of the 1970s by the physicist Freeman J. Dyson in his article "Time Without End."[12] Here, Dyson proposes nothing less than immortality within a model of the infinitely expanding Universe. He envisions other cosmic civilizations that are also in search of suitable "carriers" that will help them to survive cosmic catastrophes. Certain types of computers and robots are viewed as intelligent forms of life for this purpose, and they carry on biological evolution in a novel way.

With these technically and scientifically motivated speculations, the gray zone of humanity's present time horizon
has been reached. Clothed in technical and scientific
language, our old myths regarding time and eternity
continue to be spun out, with an ever-changing awareness of
time in the course of the socio-cultural evolution. As far as
we know, the emergence of consciousness is itself the result
of biological evolution. And the biological evolution of
living organisms far from thermal equilibrium is itself a part
of a physical-cosmic evolution, in which phase transitions
facilitate the emergence of new structures, forms, and
systems.

Since evolutionary processes are the consequences of
self-organization, they are not globally directed. Instead,
they are the products of local interactions in complex
systems under particular boundary conditions. Each step in
this process of emergence is characterized by specific interior times, and in the course of a complex evolution, these
are superimposed into a complex time hierarchy.

The interior time of a complex dynamic system refers to
its irreversible development, as revealed in the system's
possible trajectories and attractors. This dissipative and
conservative self-organization is accomplished in characteristic phase transitions at bifurcation points and is therefore
the result of temporal symmetry breaking. Put briefly, the
interior (irreversible) time of a complex dynamic system is a
measure of symmetry breaking in its dynamic development
that is described mathematically by a "time operator".

Interior time is a concept of time from the Heraclitian world—irreversible, dynamic, and constantly in a state of change. This is in contrast to the relativity theory and quantum mechanics of the Parmenides world, with their unchangeable time-symmetrical laws. There, time is merely an invariant parameter and not a dynamic operator. But a unified theory of quantum field theories and thermodynamics, from which irreversible processes would be derivable, exists only as a bare outline so far and has certainly not been theoretically and conclusively established, let alone empirically confirmed.

In Chapter 5, we learned about Penrose's expressed hope for a unified theory of (linear) quantum mechanics and (nonlinear) relativistic gravitational theory, about von Weizsäcker's logic of time (from which he proposes to derive quantum mechanics, relativity theory, and thermodynamics directly), and about Prigogine's proposal for a time operator that characterizes interior time in terms of symmetry breaking. In principle, it is conceivable that human consciousness will be able to construct and convincingly confirm such a unified cosmological theory for the elucidation of time, but it is by no means certain.

To be sure, our theories of time originate in human consciousness and are, in that sense, dependent on it. In this respect Kant's transcendental philosophy is correct, but our theories of time do nonetheless extend beyond human consciousness. For they trace the design of cosmic, physical, and biological evolution, and each of these has its characteristic interior temporal developments that facilitated the development of the human consciousness that contemplates them. In this sense, all the philosophers who presume time to be a fundamental structure of existence, antedating and independent of human existence, are correct.

These considerations also illuminate the role played by philosophy in discussions of time. Philosophy is the nucleus and the unifying element of the manifold research activities in the physical, biological, psychological, historical, cultural, and social theories of time. The philosophy of time is therefore linked to these separate sciences, but these are themselves also enmeshed with the philosophy of time. While philosophy, like all explorations, is fallible and often in need of revision, it is nonetheless a central part of the research process. In contrast to the individual sciences, it deals with their diverse aspects together, coordinates them, and examines critically the perspectives of the different disciplines. In so doing, philosophy stimulates research while creating room for reflection, expands human horizons, and averts one-sided reductionisms. Today, we understand quite a lot about the common structures shared by the concepts of time in the physical, biological, psychological, historical, cultural, social, and philosophical fields. But we are also aware of time's symmetry breaking.

1   Jaspers, *The Origin and Goal of History* (Westport, Connecticut: Greenwood Publishing Group, reprint ed. 1976).

2   Droysen, *Historik* I, ed. Leyh (Stuttgart, Germany: Frommann, 1977).

3   Weber, *Gesammelte Aufsätze zur Wissenschaftslehre* (Stuttgart, Germany. UTB, 7th ed. 1988).

4   Danto, *Analytical Philosophy of History* (Cambridge, Massachusetts: Cambridge University Press, 1965); Dray, *Laws and Explanation in History* (Westport, Connecticut: Greenwood Publishing Group, reprint ed. 1979); Hempel, *Aspects of Scientific Explanation and Other Essays in the Philosophy of Science* (New York: Free Press, 1970); Schwemmer, *Theorie der rationalen Erklärung: Zu den methodischen Grundlagen der Kulturwissenschaften* (Munich, Germany: C. H. Beck, 1976).

5   Rieter, "Quesnays Tableau Economique als Uhren-Analogie," *Studien zur Enwicklung der ökonomischen Theorie* Vol. 9, ed. Scherf (Berlin, Germany: Duncker u. H., 1990), 73.

6   Mainzer, *Thinking in Complexity: The Complex Dynamics of Matter, Mind, and Mankind* (New York: Springer-Verlag, 3rd ed. 1997), Chapter 3.

7   Goodwin, *Chaotic Economic Dynamics* (Oxford: Clarendon Press, 1990), 113.

8   Arthur, Ermoliew, and Kaniowski, "Path-dependent Processes and the Emergence of Macro-structure," *European Journal of Operational Research* 30 (1987): 294–303; Mainzer, *Thinking in Complexity: The Complex Dynamics of Matter, Mind, and Mankind* (New York: Springer-Verlag, 3rd ed. 1997), Chapter 3.

9   Nowotny, *Eigenzeit: Entstehung und Strukturierung eines Zeitgefühls* (Frankfurt, Germany: Suhrkamp, 1989).

10 Mainzer, *Computer, neue Flügel des Geistes? Die Evolution computergestützter Technik, Wissenschaft, Kultur und Philosophie* (Berlin, Germany: Walter de Gruyter, 1994), 537.

11 Moravec, *Mind Children: The Future of Robot and Human Intelligence* (Cambridge, Massachusetts: Harvard University Press, 1988); Kaiser, Matejovski, and Fedrowitz, *Kultur und Technik im 21. Jahrhundert* (Frankfurt, Germany: Campus Verlag, 1993).

12 Dyson, "Time Without End: Physics and Biology in an Open Universe," *Reviews of Modern Physics* 51 (1979): 447–460.

Further Reading

Allen, James F. "Towards a General Theory of Action and Time." *Artificial Intelligence* 23, no. 2 (1984): 123–154.

Aschoff, Jürgen (ed.). *Handbook of Behavioral Neurobiology: Biological Rhythms.* New York: Plenum Publishers, 1981.

Balslev, Anindita Niyogi. *A Study of Time in Indian Philosophy.* New Delhi: Munshiram Manoharlal Publishers Pvt. Ltd., 2nd ed. 1983.

Bergson, Henri. *Time and Free Will: An Essay on the Immediate Data of Consciousness.* Translated by F. L. Pogson. Mineola, New York: Dover Publications, 2001. (French original: *Essai sur les données immediates de la conscience.* Paris: 1889.)

Coveney, Peter, and Roger Highfield. *The Arrow of Time: A Voyage Through Science to Solve Time's Greatest Mystery.* New York: Fawcett Books, reprint ed. 1992.

Denbigh, Kenneth George. *Three Concepts of Time.* New York: Springer-Verlag, 1981.

Earman, John. *World Enough and Space-Time: Absolute Versus Relational Theories of Space and Time.* Cambridge, Massachusetts: MIT Press, 1990.

Fraser, Julius Thomas (ed.). *The Voices of Time: A Cooperative Survey of Man's View of Time as Expressed by the Sciences and by the Humanities.* New York: George Braziller, 1966.

Fraser, Julius Thomas. *Time: The Familiar Stranger.* Amherst/Boston: University of Massachusetts Press, 1987.

Friedman, William J. *The Developmental Psychology of Time.* New York: Academic Press, 1983.

Galton, Antony (ed.). *Temporal Logics and Their Applications.* New York: Academic Press, 1988.

Gould, Stephen Jay. *Time's Arrow—Time's Cycle: Myth and Metaphor in the Discovery of Geological Time.* Cambridge, Massachusetts: Harvard University Press, 1988.

Griffin, David R. (ed.). *Physics and the Ultimate Significance of Time: Bohm, Prigogine and Process Philosophy.* Albany, New York: State University of New York Press, 1985.

Gunnell, John G. *Political Philosophy of Time: Plato and the Origins of Political Vision.* Chicago: Chicago University Press, reprint ed. 1987.

Hawking, Stephen W. *A Brief History of Time: From the Big Bang to the Black Holes.* New York: Bantam Doubleday Dell Publishers, 10th anniversary ed. 1998.

Hawking, Stephen W. *The Universe in a Nutshell.* New York: Bantam Doubleday Dell Publishers, 2001.

Horwich, Paul. *Asymmetries in Time: Problems in the Philosophy of Science.* Cambridge, Massachusetts: MIT Press, 1987.

Janich, Peter. *Protophysics of Time: Constructive Foundation and the History of Time Measurement.* Vol. 30 of *Boston Studies in the Philosophy of Time.* Dordrecht, Netherlands: D. Reidel Publishing Company, 1985.

Lauer, Robert H. *Temporal Man: The Meaning and Uses of Social Time.* Westport, Connecticut: Praeger Publishing, 1981.

Mainzer, Klaus. *Symmetries of Nature: A Handbook for Philosophy of Nature and Science.* New York: Walter De Gruyter, 1996.

Mainzer, Klaus. *Thinking in Complexity: The Complex Dynamics of Matter, Mind, and Mankind.* New York: Springer-Verlag, 3rd ed. 1997.

Michon, John A., and Janet L. Jackson (eds.). *Time, Mind, and Behavior.* New York: Springer-Verlag, 1985.

Morris, Richard. *Time's Arrows: Scientific Attitudes Toward Time.* New York: Simon & Schuster, reprint ed. 1985.

Neugebauer, Otto. *History of Ancient Mathematical Astronomy*. New York: Springer-Verlag, 1975.

Newton-Smith, William H. *The Structure of Time*. London: Routledge & Kegan Paul, 1980.

Piaget, Jean. *The Child's Conception of Time*. New York: Ballantine Books, reissue ed. 1985.

Prigogine, Ilya. *From Being to Becoming: Time and Complexity in Physical Sciences*. New York: W. H. Freeman, 1981.

Prior, Arthur N. *Past, Present, and Future*. Oxford: Clarendon Press, 1967.

Quinones, Ricardo J. *The Renaissance: Discovery of Time*. Cambridge, Massachusetts: Harvard University Press, 1972.

Reichenbach, Hans. *The Philosophy of Space and Time*. Translated by Maria Reichenbach. Mineola, New York: Dover Publications, 1982.

Rescher, Nicholas, and A. J. Urquhart. *Temporal Logic*. New York: Springer-Verlag, 1971.

Toulmin, Stephen, and June Goodfield. *The Discovery of Time*. Chicago: University of Chicago Press, reprint ed. 1982.

Trivers, Howard. *Rhythm of Being: A Study of Temporality*. New York: Philosophical Library, 1985.

Van Fraassen, Bas C. *An Introduction to the Philosophy of Time and Space*. New York: Columbia University Press, Morningside ed. 1985.

Von Wright, Georg Henrik. *Time, Change, and Contradiction: The Twenty-second Arthur Stanley Eddington Memorial Lecture, Delivered at Cambridge University, 1 November 1968.* Cambridge, Massachusetts: Cambridge University Press, 1968.

Wessman, Alden E., and Bernard S. Gorman. *The Personal Experience of Time.* New York: Plenum Publishers, 1977.

Whitrow, G. J. *Natural Philosophy of Time.* New York: Oxford University Press, 2nd ed. 1982.

Winfree, Arthur T. *The Geometry of Biological Time.* New York: Springer-Verlag, 2nd ed. 2001.

Winston, Gordon C. *The Timing of Economic Activities: Firms, Households, and Markets in Time-Specific Analysis.* Cambridge, Massachusetts: Cambridge University Press, 1982.

Zeman, Jiří (ed.). *Time in Science and Philosophy: An International Study of Some Current Problems.* Amsterdam/New York: Elsevier Science 1971.

Index